Lecture Notes in Biomathematics

Managing Editor: S. Levin

17

Takeo Maruyama

Stochastic Problems
in Population Genetics

Springer-Verlag
Berlin Heidelberg New York

Lecture Notes in Biomathematics

Lecture Notes in Biomathematics

Managing Editor: S. Levin

17

Takeo Maruyama

Stochastic Problems in Population Genetics

Springer-Verlag
Berlin Heidelberg New York 1977

Author

Dr. Takeo Maruyama
National Institute of Genetics
Yata 1, 111 Mishima
Sizuoka-Ken
411 Japan

Library of Congress Cataloging in Publication Data

Maruyama, Takeo, 1936-
    Stochastic problems in population genetics.

    (Lecture notes in biomathematics ; 17)
    Bibliography: p.
    Includes index.
    1. Population genetics--Mathematical models.
2. Stochastic processes. I. Title. II. Series.
QH455.M36   575.1   77-24644

AMS Subject Classifications (1970): 92-02, 92 A10, 92 A15

ISBN 3-540-08349-9 Springer-Verlag Berlin Heidelberg New York
ISBN 0-387-08349-9 Springer-Verlag New York Heidelberg Berlin

Printing and binding: Beltz Offsetdruck, Hemsbach/Bergstr.
2145/3140-543210

PREFACE

These are notes based on courses in Theoretical Population
Genetics given at the University of Texas at Houston during the winter
quarter, 1974, and at the University of Wisconsin during the fall
semester, 1976. These notes explore problems of population genetics
and evolution involving stochastic processes. Biological models and
various mathematical techniques are discussed. Special emphasis is
given to the diffusion method and an attempt is made to emphasize the
underlying unity of various problems based on the Kolmogorov backward
equation. A particular effort was made to make the subject accessible
to biology students who are not familiar with stochastic processes.
The references are not exhaustive but were chosen to provide a starting
point for the reader interested in pursuing the subject further.

Acknowledgement   I would like to use this opportunity to
express my thanks to Drs. J. F. Crow, M. Nei and W. J. Schull for
their hospitality during my stays at their universities. I am indebted
to Dr. M. Kimura for his continuous encouragement. My thanks also go
to the small but resolute groups of students, visitors and colleagues
whose enthusiasm was a great source of encouragement. I am especially
obliged to Dr. Martin Curie-Cohen and Dr. Crow for reading a large part of the
manuscript and making many valuable comments. Special gratitude is
expressed to Miss Sumiko Imamiya for her patience and endurance and
for her efficient preparation of the manuscript.
Finally I must thank the agencies which enabled me to work at
these universities: Ministry of Education (Japan); National Institute
of Health (U. S. A.) GM-19513 and GM-20293 through University of Texas,
GM-22038 through University of Wisconsin.

TABLE OF CONTENTS

CHAPTER 1

ORIENTATION

It is the purpose of this chapter to acquaint the reader with certain mathematical notions and theorems used throughout the book. To do this, we will begin not with a biological model, but with a random walk process on a set of integers, which is one of the simplest Markov processes. Because of its mathematical simplicity, this process is an ideal starting point, enabling us to learn all the essential concepts and methods related to stochastic problems.

1.1 Discrete space, continuous time, random walk

Consider a random walk on a set of integers, $\{1, 2, 3, \cdots, n\}$, with equal probability of moving to the right and to the left, and assume that the waiting time obeys the negative exponential law, e^{-t}, i.e. the probability that a jump does not occur in a time interval $(0, t)$ is equal to e^{-t}. Assume further that when the walk reaches 0 or $n+1$, it is absorbed. Denote by X_t the position at time t, and let

$$p_{ij}^{(t)} = \text{prob.} \left\{ X_t = i \mid X_0 = j \right\}$$

which is the probability that the position at time t is i given that the process starts from j at time 0. It is obvious that

$$p_{ij}^{(0)} = \delta_{ij} = 0 \qquad i \neq j$$
$$= \delta_{ij} = 1 \qquad i = j.$$

For small Δt, we have the following relationship between $p_{ij}^{(t+\Delta t)}$ and $p_{ij}^{(t)}$,

$$(1.1) \quad p_{ij}^{(t+\Delta t)} = (1-\Delta t)p_{ij}^{(t)} + \frac{\Delta t}{2}\left(p_{i-1,j}^{(t)} + p_{i+1,j}^{(t)} \right) + o(\Delta t) .$$

The first term on the right side is the probability that the position at time t is i and that no change in the position occurs in the following time interval Δt. The second term gives the probability that the position is not at i at time t, but that it moves to i by a single jump in the time interval Δt. The last term accounts for the higher order events which were omitted from the first two terms. The symbol $o(\Delta t)$ means

$$o(\Delta t)/\Delta t \to 0$$

as $\Delta t \to 0$. Substracting $p_{ij}^{(t)}$ from both sides of (1.1), dividing by Δt, and letting $\Delta t \to 0$, we have

(1.2)
$$\frac{dp_{ij}^{(t)}}{dt} = \frac{1}{2}\left\{p_{i-1,j}^{(t)} - 2p_{ij}^{(t)} + p_{i+1,j}^{(t)}\right\}.$$

Using matrix and vector notation, this equation can be rewritten as

(1.3)
$$\frac{dP^{(t)}}{dt} = \begin{bmatrix} -1 & \frac{1}{2} & & & \mathbf{O} \\ \frac{1}{2} & -1 & \cdot & & \\ & \cdot & \cdot & \cdot & \frac{1}{2} \\ & & \cdot & \cdot & \\ \mathbf{O} & & \frac{1}{2} & -1 \end{bmatrix}_{n \times n} P^{(t)}$$

where $P^{(t)}$ is the matrix consisting of $p_{ij}^{(t)}$, i.e., $P^{(t)} = [p_{ij}^{(t)}]$, and

$$\frac{dP^{(t)}}{dt} = \left[\frac{dp_{ij}^{(t)}}{dt}\right].$$

Similarly, with respect to the index j of $p_{ij}^{(t)}$, we have

$$p_{ij}^{(t+\Delta t)} = (1-\Delta t)p_{ij}^{(t)} + \frac{\Delta t}{2}(p_{i,j-1}^{(t)} + p_{i,j+1}^{(t)}) + o(\Delta t)$$

and thus

(1.4)
$$\frac{dp_{ij}^{(t)}}{dt} = \frac{1}{2}\left\{p_{i,j-1}^{(t)} - 2p_{ij}^{(t)} + p_{i,j+1}^{(t)}\right\}.$$

This can be written as

$$(1.5) \qquad \frac{dP^{(t)}}{dt} = P^{(t)} \begin{bmatrix} -1 & \frac{1}{2} & & & \text{\Large O} \\ \frac{1}{2} & -1 & \cdot & & \\ & \cdot & \cdot & \cdot & \\ & & \cdot & \cdot & \frac{1}{2} \\ \text{\Large O} & & & \frac{1}{2} & -1 \end{bmatrix}_{n \times n}.$$

The system (1.3) and (1.5) are called the Kolmogorov "forward" and "backward" equations, respectively. The two systems are the transpositions of each other, but appear to be quite similar. We will briefly examine the differences between the two with respect to our particular interests. Note that in (1.3) the right-multiplication of a vector on $P^{(t)}$ is possible, while in (1.5) the left-multiplication of a vector on $P^{(t)}$ is possible. Now let $w = (w_1, w_2, \ldots, w_n)$ be a row vector consisting n constants. Then

$$wP^{(t)} = \left(\sum_k w_k P_{k1}^{(t)}, \sum_k w_k P_{k2}^{(t)}, \ldots, \sum_k w_k P_{kn}^{(t)} \right)$$

is the vector consisting of the averages of the $P_{ki}^{(t)}$ averaged over k and weighted by w_k. Namely the i-th element of the vector is the expectation of the quantity given by a random variable which assigns w_k to the state k, provided that the process starts from i. For example, if $w = (1, 1, \cdots, 1)$, the $wP^{(t)}$ is simply $\sum_k P_{ki}^{(t)}$ and therefore it gives the probability that the process stays in the state space (1, 2, $\cdots$, n) until time t. If we let $v^{(t)} \equiv (v_1^{(t)}, v_2^{(t)}, \ldots, v_n^{(t)}) \equiv wP^{(t)}$, and multiply both sides of (1.5) by the vector w, we have

$$(1.6) \qquad \frac{dv^{(t)}}{dt} = v^{(t)} A$$

where A is the matrix appearing at the right of equation (1.5). Equation (1.6) is also called the Kolmogorov backward equation. Hence the appropriate solution of (1.6) gives the time dependent expectations of the quantity given by the weight vector w. Since $p_{ij}^{(0)} = \delta_{ij}$,

$$v^{(0)} = w,$$

which is the initial condition. Equation (1.6) can be solved by the method for an ordinary differential equation of type

$$f'(t) = af(t)$$

of which the appropriate solution is

$$f(t) = f(0)e^{at} .$$

In analogy with this, we have

(1.7)
$$v^{(t)} = we^{tA}$$

with

(1.8)
$$e^{tA} = I + tA + \frac{t^2}{2!}A^2 + \frac{t^3}{3!}A^3 + \cdots$$

where I is the identity matrix of the same order as A. Thus the solution of (1.5) is

(1.9)
$$p^{(t)} = p^{(0)}e^{tA} = e^{tA}$$

(noting $p^{(0)} = I$). We can heuristically show that the matrix $p^{(t)}$ of (1.9), which is defined in (1.8), is indeed the solution of the problem.
We shall now prove the above statement. Since matrix A has positive and negative elements, the probabilistic meaning is not clear. But note that A can be decomposed into two parts:

(1.10)
$$A = S - I$$

where I is the identity matrix and S is

(1.11)
$$\begin{pmatrix} 0 & \frac{1}{2} & & & O \\ \frac{1}{2} & 0 & \cdot & & \\ & \cdot & \cdot & \cdot & \frac{1}{2} \\ & & \cdot & \cdot & \cdot \\ O & & \frac{1}{2} & 0 \end{pmatrix} .$$

Now note that S is the one-step transition probability matrix, given that the position is changed, and further that $S^i = \underbrace{SS \cdots S}_{i \text{ times}}$ is the i-step transition matrix. Returning to (1.9), rewrite it as

(1.12)
$$p^{(t)} = e^{tA} = e^{-tI}e^{tS}.$$

According to definition (1.8), these two terms have the following series expansions,

(1.13) $\quad e^{-tI} = I + (-tI) + \frac{1}{2!}(-tI)^2 + \cdots = e^{-t}I$

and

(1.14) $\quad e^{tS} = I + tS + \frac{1}{2!}t^2S^2 + \cdots .$

Therefore, upon substitution of (1.13) and (1.14) into (1.12), we have

(1.15) $\quad P^{(t)} = e^{-t}\left\{ I + tS + \frac{1}{2!}t^2S^2 + \cdots + \right\} = e^{-t}\sum_{k=0}^{\infty} \frac{t^k}{k!}S^k .$

The last formula in (1.15) is certainly the solution which we seek. Why? The coefficient of the k-th term

$$\frac{e^{-t}t^k}{k!}$$

is the probability that the process makes exactly k jumps in the time interval (0, t), and the elements s_{ij} of matrix

$$\frac{e^{-t}t^k}{k!}S^k$$

is the probability that exactly k jumps occur in (0, t) and that the position at time t is i, given that it is j at t = 0. Thus the sum of all such possibilities, which is given by (1.15), must be the correct solution.

We can simplify formula (1.15). Suppose there are two mutually orthonormal matrices, L and R, such that

(1.16) $\quad LSR = \Lambda$

where Λ is a diagonal matrix consisting of λ_1, λ_2, ..., and "orthonormal" means that LR = I. Applying the orthogonality of L and R,

$S = L^{-1}\Lambda R^{-1}$

$S^2 = (L^{-1}\Lambda R^{-1})(L^{-1}\Lambda R^{-1}) = L^{-1}\Lambda(R^{-1}L^{-1})\Lambda R^{-1} = L^{-1}\Lambda^2 R^{-1}$

$S^3 = (L^{-1}\Lambda R^{-1})(L^{-1}\Lambda R^{-1})(L^{-1}\Lambda R^{-1})$

$\quad = L^{-1}\Lambda(R^{-1}L^{-1})\Lambda(R^{-1}L^{-1})\Lambda R^{-1} = L^{-1}\Lambda^3 R^{-1}$

$\vdots$

$S^k = L^{-1}\Lambda^k R^{-1} .$

Therefore (1.15) can be rewritten as

(1.17) $$P^{(t)} = e^{-t}{}_L{}^{-1} \left\{ I + t\Lambda + \frac{t^2}{2!} \Lambda^2 + \cdots \right\} R^{-1}.$$

Note that

(1.18) $$\Lambda^k = \begin{pmatrix} \lambda_1^k & & & & \mathbf{0} \\ & \lambda_2^k & & & \\ & & \lambda_3^k & & \\ & & & & \ddots \\ \mathbf{0} & & & & \end{pmatrix}.$$

Therefore

(1.19) $$I + t\Lambda + \frac{t^2}{2!} \Lambda^2 + \cdots = \begin{pmatrix} e^{\lambda_1 t} & & & & \mathbf{0} \\ & e^{\lambda_2 t} & & & \\ & & e^{\lambda_3 t} & & \\ & & & & \ddots \\ \mathbf{0} & & & & \end{pmatrix}.$$

Denote the above diagonal matrix by $\Lambda^{(t)}$. Then using these notations,

(1.20) $$P^{(t)} = e^{-t}{}_L{}^{-1} \Lambda^{(t)} {}_R{}^{-1}.$$

As a candidate for R and L defined in (1.16) for matrix S given in (1.11), let us examine the following matrix

(1.21) $$E = \frac{\sqrt{2}}{\sqrt{n+1}} \begin{bmatrix} \sin \frac{\pi}{n+1} & \sin \frac{2\pi}{n+1} & \cdots & \sin \frac{n\pi}{n+1} \\ \sin \frac{2\pi}{n+1} & \sin \frac{2\cdot 2\pi}{n+1} & \cdots & \sin \frac{2n\pi}{n+1} \\ \vdots & \vdots & & \vdots \\ \sin \frac{n\pi}{n+1} & \sin \frac{2n\pi}{n+1} & \cdots & \sin \frac{n^2\pi}{n+1} \end{bmatrix}.$$

First note the orthogonality,

$$EE = I$$

and also that

$$(1.22) \qquad ES = \frac{\sqrt{2}}{\sqrt{n+1}} \left[\cos \frac{\pi}{n+1} E_1, \ \cos \frac{2\pi}{n+1} E_2, \ \ldots, \ \cos \frac{n\pi}{n+1} E_{n-1} \right]$$

where E_k is the k-th column of matrix E, i.e., $E = [E_1, E_2, \ldots, E_n]$. If we multiply both sides of S by E, we have

$$(1.23) \qquad ESE = \begin{bmatrix} \cos \frac{\pi}{n+1} & & & & \text{\Large 0} \\ & \cos \frac{2\pi}{n+1} & & & \\ & & \cos \frac{3\pi}{n+1} & & \\ & & & & \ddots \\ \text{\Large 0} & & & & \end{bmatrix}.$$

Therefore

$$(1.24) \qquad P^{(t)} = E \begin{bmatrix} e^{(\cos \frac{\pi}{n+1} - 1)t} & & & \text{\Large 0} \\ & e^{(\cos \frac{2\pi}{n+1} - 1)t} & & \\ & & \ddots & \\ \text{\Large 0} & & & \end{bmatrix} E$$

where E is defined in (1.21).

Problem 1. Write each element $p_{ij}^{(t)}$ of $P^{(t)}$ explicitly in terms of the sine function and $\exp \left\{ (\cos \frac{k\pi}{n+1}) - 1)t \right\}$.

Starting from equation (1.5), we have obtained solution (1.12). We shall now prove the converse: starting from (1.9) we can recover (1.3) or (1.5).

Problem 2. Prove $e^{(t+s)A} = e^{tA} e^{sA} = e^{sA} e^{tA}$.

Assuming Problem 2 has been proven, let us calculate

$$\frac{1}{\Delta t} \left\{ P^{(t+\Delta t)} - P^{(t)} \right\}$$

$$(1.25) \qquad = \frac{1}{\Delta t} \left\{ e^{(t+\Delta t)A} - e^{tA} \right\} = \frac{1}{\Delta t} e^{tA} (e^{\Delta t A} - I).$$

Likewise

$$(1.26) \qquad \frac{1}{\Delta t} \left\{ e^{(\Delta t + t)A} - e^{tA} \right\} = \frac{1}{\Delta t} (e^{\Delta t A} - I) e^{tA}.$$

Now

$$\frac{1}{\Delta t} (e^{\Delta t A} - I) = \frac{1}{\Delta t} \left\{ \Delta t A + \frac{(\Delta t)^2}{2!} A^2 + \cdots \right\}.$$

Therefore, noting that $\frac{1}{\Delta t} \frac{(\Delta t)^2}{2!} A^2$ and the higher order terms vanish as $\Delta t \to 0$,

$$(1.27) \qquad \lim_{\Delta t \to 0} \frac{1}{\Delta t} (e^{\Delta t A} - I) = A .$$

Therefore we have proven

$$(1.28) \qquad \frac{dP^{(t)}}{dt} = P^{(t)} A$$

and

$$(1.29) \qquad \frac{dP^{(t)}}{dt} = A P^{(t)} .$$

We shall examine more general cases. Let $S = [s_{ij}]$ be any one-step transition probability matrix. Then

$$(1.30) \qquad P^{(t)} = e^{t(S-I)} = e^{-t} \sum_{k=0}^{\infty} \frac{t^k}{k!} S^k$$

is the transition probability matrix of a process whose waiting time obeys the negative exponential law e^{-t}.

Problem 3. Prove the above statement heuristically.

Problem 4. Derive

$$(1.31) \qquad \frac{dP^{(t)}}{dt} = (S - I) P^{(t)}$$

by an analogous argument used in deriving (1.3) or (1.5).

Problem 5. Suppose that, when the position is changed, it moves one step in one direction with probability x and to the other direction with probability $y = 1-x$, where not necessarily $x = y = 1/2$. Derive the differential equation for $P^{(t)}$ and obtain the solution.

Problem 6, (Continuation of Problem 5). Suppose that $x+y < 1$ and the position remains unchanged after a jump with probability $z = 1-x-y$. Do the same questions as Problem 5.

There are two generalizations of the processes discussed above. One is that a jump occurs in the time interval Δt with probability

$$\theta \Delta t + o(\Delta t)$$

instead of $\Delta t + o(\Delta t)$. Then the equation corresponding to (1.1) is

$$(1.32) \qquad \frac{dp_{ij}^{(t)}}{dt} = \theta \left\{ \frac{1}{2} p_{i-1,j}^{(t)} - p_{ij}^{(t)} + \frac{1}{2} p_{i+1,j}^{(t)} \right\}$$

or, in the matrix form,

$$(1.33) \qquad \frac{dP^{(t)}}{dt} = \theta \begin{bmatrix} -1 & \frac{1}{2} & & & & \huge 0 \\ \frac{1}{2} & -1 & \frac{1}{2} & & & \\ & \cdot & \cdot & \cdot & & \\ & & \cdot & \cdot & \cdot & \frac{1}{2} \\ \huge 0 & & & \cdot & \cdot & \cdot \\ & & & \frac{1}{2} & -1 \end{bmatrix} P^{(t)}.$$

Therefore the solution is given by

$$(1.34) \qquad P^{(t)} = P^{(0)} e^{\theta t A} = e^{\theta t A}$$

where A is the same matrix as before. Thus the solution of the new problem is the same as the previous one, except that time t is changed to θt.

The second generalization is that we may assume that a jump of more than one step can occur and that its probability is

$$\theta_k \Delta t + o(\Delta t).$$

This means that in time interval Δt, a change of k-steps to the right or the left occurs with probability $\theta_k \Delta t + o(\Delta t)$. Then the transition probability $p_{ij}^{(t)}$ satisfies

$$p_{ij}^{(t+\Delta t)} = (1 - \Delta t \sum_k \theta_k) p_{ij}^{(t)} + \frac{\Delta t}{2} \left\{ \sum_{k \neq 0} \theta_k \left(p_{i-k,j}^{(t)} + p_{i+k,j}^{(t)} \right) \right\} + o(\Delta t)$$

and

$$(1.35) \qquad \frac{dp_{ij}^{(t)}}{dt} = \sum_{k \neq 0} \theta_k \left\{ \frac{1}{2} p_{i-k,j}^{(t)} - p_{ij}^{(t)} + \frac{1}{2} p_{i+k,j}^{(t)} \right\} - \theta_0 p_{ij}^{(t)} .$$

We can rewrite these equations as

$$(1.36) \qquad \frac{dP^{(t)}}{dt} = \left[(1 - \sum_{k \neq 0} \theta_k) I + \frac{1}{2} \sum_{k \neq 0} \theta_k S_k \right] P^{(t)}$$

where I is the identity matrix and

$$
\begin{array}{c}
\text{k-th}\!\downarrow\!\text{column} \\
S_k = \\
\text{k-th row} \to
\end{array}
\begin{bmatrix}
0 & 0 & . & 1 & & \text{\large O} \\
0 & 0 & 0 & . & 1 & \\
. & 0 & . & . & . & 1 \\
1 & . & . & . & . & . \\
& 1 & . & . & . & 0 \\
\text{\large O} & . & 1 & . & 0 & 0
\end{bmatrix}
$$

namely the k-th diagonal above the main diagonal and the k-th below it consist of 1's and all the other elements are zeros.

The system of differential equations (1.36) has the same set of eigenvectors and a similar set of eigenvalues as given in (1.24). The solution of (1.36) is

$$
p^{(t)} = E
\begin{bmatrix}
e^{(\Sigma\theta_k \cos\frac{\pi k}{n} - 1)t} & & \text{\large O} \\
& e^{(\Sigma\theta_k \cos\frac{2\pi k}{n} - 1)t} & \\
\text{\large O} & & . \\
& & . \;.
\end{bmatrix}
E
$$

where E is the matrix given in (1.21).

1.2 Discrete space, discrete time

There is another type of process which is similar to the one discussed above. We may assume that a jump occurs with certain probability at only discrete times, t = 1, 2, $\cdots$. For simplicity, assume that a jump of more than one step does not occur. Let $\theta/2$ be the probability of moving to the right (or left) for the discrete times t = 1, 2, 3, $\cdots$. Then

(1.37)
$$
\theta
\begin{bmatrix}
0 & \frac{1}{2} & 0 & & \text{\large O} \\
\frac{1}{2} & 0 & \frac{1}{2} & . & \\
& . & . & . & . \\
\text{\large O} & . & . & . & .
\end{bmatrix}
+ (1 - \theta)I
$$

gives the transition probabilities for t = 1. Therefore the transition probabilities at time t are given by

(1.38)
$$
p^{(t)} = [\theta S + (1-\theta)I]^{t} \qquad t = 1, 2, \cdots,
$$

where S is the matrix appearing in (1.37). Equation (1.38) corresponds to equation (1.15) of the continuous time case. The eigenvectors of matrices S and $\theta S + (1-\theta)T$ are the same, and eigenvalues are

$$\lambda'_k = \theta\lambda_k + (1 - \theta) = 1 + \theta(\cos \frac{k\pi}{n+1} - 1) ,$$

where $\lambda_k = \cos \frac{k\pi}{n+1}$ is an eigenvalue of S. As a generalization of this, the following theorem of algebra seems useful.

Theorem. Let A be any square matrix, and let $\{\lambda_1, \lambda_2, ...\}$ and $\{v_1, v_2, ...\}$ be the eigenvalues and eigenvectors of A. Let p(x) be any polynomial in x. Then matrix

$$p(A)$$

has the same set of the eigenvectors as A and the corresponding eigenvalues $\{p(\lambda_k)\}$.

As before, we can write $P^{(t)}$ of (1.38) in terms of its spectrum:

(1.39)
$$P^{(t)} = E \begin{bmatrix} \left\{1+\theta(\cos \frac{\pi}{n+1} - 1)\right\}^t & & O \\ & \left\{1+\theta(\cos \frac{2\pi}{n+1} - 1)\right\}^t & \\ O & & \ddots \end{bmatrix} E$$

where E is the matrix given in (1.21).

Note that the solution given in (1.39) of the discrete time parameter case and the solution of the continuous time parameter case given in (1.24) are quite similar. Particularly, if the parameter θ is small,

(1.40)
$$1 + \theta(\cos \frac{k\pi}{n+1} - 1) \approx e^{\theta(\cos \frac{k\pi}{n+1} - 1)}$$

and

(1.41)
$$\left\{1 + \theta(\cos \frac{k\pi}{n+1} - 1)\right\}^t \approx e^{\theta(\cos \frac{k\pi}{n+1} - 1)t} .$$

Therefore if we change the unit of time and the parameter θ in such a way so that the product of the two remains unchanged, (1.39) converges to the solution (1.24) of the continuous time parameter.

1.3 Circular space, continuous time

An interesting modification of the above model of continuous time parameter is to assume that state n is identified with 0, and thus the state space consists of n point $\{0, 1, 2, \cdots, n-1\}$, and is circular. This is a space which is finite but without end. Using the same argument used for the derivation of equations (1.3) and (1.5), we have

$$
(1.42) \qquad \frac{dp^{(t)}}{dt} = \theta
\begin{bmatrix}
-1 & \frac{1}{2} & 0 & 0 \cdots 0 & \frac{1}{2} \\
\frac{1}{2} & -1 & \frac{1}{2} & 0 \cdots & 0 \\
0 & & & & 0 \\
\vdots & 0 & & & \frac{1}{2} \\
\frac{1}{2} & 0 & \cdots\cdots & \frac{1}{2} & -1
\end{bmatrix}
P^{(t)}
$$

and

$$
(1.43) \qquad \frac{dp^{(t)}}{dt} = \theta P^{(t)} A
$$

where A is the same matrix that appears on the right side of (1.42). Let C be the matrix

$$
(1.44) \qquad \frac{\sqrt{2}}{\sqrt{N}}
\begin{bmatrix}
\frac{1}{2} & 1 & 1 & \cdots & 1 \\
\frac{1}{2} & \cos\frac{2\pi}{n} & \cos\frac{3\cdot 2\pi}{n} & \cdots & \cos\frac{(n-1)2\pi}{n} \\
\frac{1}{2} & \cos\frac{2\cdot 2\pi}{n} & \cos\frac{3\cdot 2\cdot 2\pi}{n} & \cdots & \cos\frac{(n-1)2\cdot 2\pi}{n} \\
\frac{1}{2} & \cos\frac{3\cdot 2\pi}{n} & \cos\frac{3\cdot 3\cdot 2\pi}{n} & & \vdots \\
\vdots & \vdots & \vdots & & \vdots \\
\frac{1}{2} & \cos\frac{(n-1)2\pi}{n} & \cos\frac{3(n-1)2\pi}{n} & \cdots & \cos\frac{(n-1)^2 2\pi}{n}
\end{bmatrix}.
$$

Note that

$$
(1.45) \quad C^*
\begin{bmatrix}
-1 & \frac{1}{2} & & & \frac{1}{2} \\
\frac{1}{2} & -1 & \frac{1}{2} & & \\
 & \ddots & \ddots & \ddots & \\
 & & \ddots & \ddots & \frac{1}{2} \\
\frac{1}{2} & & & \frac{1}{2} & 1
\end{bmatrix}
C =
\begin{bmatrix}
0 & & & \\
 & (\cos\frac{2\pi}{n} - 1) & & \\
 & & (\cos\frac{2\cdot 2\pi}{n} - 1) & \\
 & & & \ddots
\end{bmatrix}
$$

where C* is the transpose of C. Thus according to formula (1.24), the solution of equation (1.42) is

$$
(1.46) \qquad P^{(t)} \; = \; C* \begin{bmatrix} 1 & & & \\ & e^{\theta(\cos\frac{2\pi}{n} - 1)t} & & \mathbf{0} \\ & & e^{\theta(\cos\frac{2\cdot2\pi}{n} - 1)t} & \\ & \mathbf{0} & & \ddots \end{bmatrix} C
$$

where C is the matrix given in (1.44)

We now compare the solution given in (1.39) and (1.46). The former is the solution for a situation where 0 and n+1 are absorbing states, or absorbing boundaries, where the process stops moving, and the latter is for a circular situation without end.

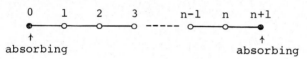

Fig. 1.1 Linear space.

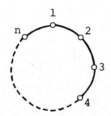

Fig. 1.2 Circular space.

An individual element of the solution given in the matrix form in (1.39) is

$$
(1.47) \qquad p_{ij}^{(t)} \; = \; \frac{2}{n} \sum_{k=1}^{n-1} e^{\theta(\cos\frac{k\pi}{n} - 1)t} \sin\frac{k\pi i}{n} \sin\frac{k\pi j}{n}
$$

and an individual element of (1.46) is

$$
(1.48) \qquad p_{ij}^{(t)} \; = \; \frac{2}{n} \sum_{k=0}^{n-1} e^{\theta(\cos\frac{2\pi k}{n} - 1)t} \delta_k \cos\frac{k2\pi i}{n} \cos\frac{k2\pi j}{n}
$$

where $\delta_0 = 1/2$ and $\delta_k = 1$ for $k > 0$. Note that as N gets indefinitely large, formulae (1.47) and (1.48) become, respectively,

$$
(1.49) \qquad p_{ij}^{(t)} \; = \; 2 \int_0^1 e^{\theta(\cos\pi x - 1)t} \sin i\pi x \sin j\pi x \; dx
$$

and

(1.50) $\qquad p_{ij}^{(t)} = 2 \int_0^1 e^{\theta(\cos \pi x - 1)t} \cos i\pi x \cos j\pi x \, dx.$

Since

$$\sin\alpha \, \sin\beta = \frac{1}{2} \left[\cos(\alpha - \beta) - \cos(\alpha + \beta) \right]$$

and

$$\cos\alpha \, \cos\beta = \frac{1}{2} \left[\cos(\alpha - \beta) + \cos(\alpha + \beta) \right],$$

substitution of these identities in (1.49) and (1.50) yields

(1.49') $\qquad p_{ij}^{(t)} = \int_0^1 e^{\theta(\cos \pi x - 1)t} [\cos \pi(i-j)x] \, dx$

$$- \int_0^1 e^{\theta(\cos \pi x - 1)t} [\cos \pi(i+j)x] \, dx$$

and

(1.50') $\qquad p_{ij}^{(t)} = \int_0^1 e^{\theta(\cos \pi x - 1)t} [\cos \pi(i-j)x] \, dx$

$$+ \int_0^1 e^{\theta(\cos \pi x - 1)t} [\cos \pi(i+j)x] \, dx.$$

Note that for every integrable function $f(x)$

$$\lim_{n \to \infty} \int_0^1 f(x) \sin nx \, dx = \int_0^1 f(x) \cos nx \, dx = 0 \;.$$

Therefore as $(i+j)$ becomes large while $(i-j)$ is kept bounded, formulae (1.49') and (1.50') converge to the identical formula

$$p_{ij}^{(t)} = \int_0^1 e^{\theta(\cos \pi x - 1)t} \cos(i-j)\pi x \, dx \;.$$

So far we have mainly dealt with the solution $P^{(t)}$ of (1.5), but in population genetics the quantity given by the vector v of (1.6) is often important, particularly the integral or sum of $v_i^{(t)}$ over the whole duration of the process. Now let

$$z_i = \int_0^\infty v_i^{(t)} \, dt$$

for the case of a continuous time parameter, and

$$z_i = \sum_{t=0}^\infty v_i^{(t)}$$

for the discrete time parameter case.

Since

$$\int_0^\infty \frac{\partial v_i^{(t)}}{\partial t}\, dt = v_i^{(\infty)} - v_i^{(0)} = v_i^{(\infty)} - w_i ,$$

It follows from (1.6) that

$$v^{(\infty)} - w = zA$$

where $z = (z_1, z_2, \cdots)$. If $v^{(\infty)}$ is a zero vector, that is $v_i^{(\infty)} = 0$ for all i, the solution of the above equation is

$$z = -wA^{-1} .$$

In the case of a discrete time parameter, $P^{(n+1)} = P^{(n)}A$ for $n = 1, 2,$ $\cdots$, and therefore $v^{(n+1)} = v^{(n)}A = wA^n$. Hence

(1.51)
$$z = \sum_{n=0}^\infty v^{(n)} = w \sum_{n=0}^\infty A^n.$$

Note that

$$\sum_{n=0}^\infty A^n = (I - A)^{-1} ,$$

and therefore

(1.52)
$$z = w(I - A)^{-1}.$$

Formulae (1.51) and (1.52) give the sums of the quantities $v_i^{(t)}$ over the whole duration as a function of the initial state. The two formulae are analogous to each other, but later we shall encounter still another type of formulae which are analogous to those.

1.4 Continuous space, continuous time

We have dealt with processes whose state spaces consist of integers. Now we derive a process on a whole interval (a, b) where a and b can be either finite or infinite. There are a number of ways of doing this depending upon mathematical sophistication. We shall take an approach which is a continuation of the problem discussed above. A physical interpretation of the process satisfying differential equation (1.33) is that a particle moves from one state to the next state to the right or the left with probability

$$\theta \Delta t + o(\Delta t)$$

in the time interval Δt, and the distance between a pair of adjacent states is one. Now suppose that we insert an additional state point between each pair of adjacent points, so that the distance between the new adjacent points is 1/2. Further suppose that the probability that a jump occurs in Δt is increased by a factor $4 = 2^2$, i.e.,

$$2^2 \theta \Delta t + o(\Delta t) .$$

Continue this procedure of inserting new points between adjacent pairs, in such a way so that Δt and the distance between points Δx satisfies

$$(\Delta x)^2 \cdot \Delta t = \theta$$

Let $f(x)$ be any function defined on a set of points $\{\cdots \frac{1}{n}, \frac{2}{n}, \frac{3}{n} \cdots\}$. Then the operation

$$n^2 \theta \begin{bmatrix} -1 & \frac{1}{2} & & & \bigcirc \\ \frac{1}{2} & -1 & \frac{1}{2} & & \\ & \cdot & \cdot & \cdot & \\ & & \cdot & \cdot & \cdot \\ \bigcirc & & & \frac{1}{2} & -1 \end{bmatrix} f(x)$$

can be regarded as a second derivative, i.e.,

$$n^2 \theta \begin{bmatrix} -1 & \frac{1}{2} & & & \bigcirc \\ \frac{1}{2} & \cdot & \cdot & \cdot & \\ & \cdot & \cdot & \cdot & \frac{1}{2} \\ \bigcirc & & \frac{1}{2} & -1 \end{bmatrix} \begin{bmatrix} f(x_1) \\ f(x_2) \\ \vdots \\ f(x_n) \end{bmatrix}$$

$$= \theta \begin{bmatrix} \dfrac{f(x_1) - 2f(x_2) + f(x_3)}{(\frac{1}{n})^2} \\[2em] \dfrac{f(x_2) - 2f(x_3) + f(x_4)}{(\frac{1}{n})^2} \\[2em] \dfrac{f(x_3) - 2f(x_4) + f(x_5)}{(\frac{1}{n})^2} \\[2em] \vdots \end{bmatrix}$$

$$\approx \theta \begin{bmatrix} f''(x_1) \\ f''(x_2) \\ \cdot \\ \cdot \\ \cdot \end{bmatrix}.$$

For the continuous state space, if we let $p(t, x, y)$ be the transition probability that $X_t = y$ at time t, given that $X_0 = x$, and apply the operation above, we have in the limiting case of $n \to \infty$,

$$\frac{\partial p(t, x, y)}{t} = \theta \frac{\partial^2 p(t, x, y)}{y^2}$$

and similarly

$$\frac{\partial p(t, x, y)}{\partial t} = \theta \frac{\partial^2 p(t, x, y)}{\partial x^2}.$$

1.5 Markov Process

A random variable X_t with time parameter $t \geq 0$ is said to be Markovian, if for arbitrary

$$s_1 < s_2 < \ldots < s_m < t_1 < t_2 < \ldots < t_n,$$

$$\text{Prob } \{X_{t_1} = x_1, X_{t_2} = x_2, \ldots, X_{t_n} = x_n \mid X_{s_m} = y_m\}$$

$$= \text{Prob } \{X_{t_1} = x_1, X_{t_2} = x_2, \ldots, X_{t_n} = x_n \mid$$

$$X_{s_1} = y_1, X_{s_2} = y_2, \ldots, X_{s_m} = y_m\}.$$

The significance of this condition is as follows: If we know the value assumed by the random variable X_t at a certain moment of time, then any information concerning the development of the random variable before this moment is without influence on our knowledge of its development in the future.

The function

$$p(t, x, y) \equiv \text{Prob } \{X_t = y \mid X_0 = x\}$$

is called the transition probability density of the random variable X_t. The function $p(t, x, y)$ possesses the following properties:

(1.53)
$$p(t+s, x, y) = \int p(s, x, \xi)p(t, \xi, y)d\xi,$$

known as the Chapman-Kolmogorov equation. For all x, p(t, x, y) is continuous to the left in t. This is equivalent to

(1.54) $$\lim_{\Delta t \to 0} \int p(\Delta t, x, y) f(y) dy = f(x)$$

where f(x) is an arbitrary function differentiable up to the second order at x.

The transition density function together with some postulates concerning the regularity of the trajectories determines all the properties of X_t.

Assume that the probability density p(t, x, y) on an interval $J = [\gamma_0, \gamma_1]$ has, for x $\in$ (γ_0, γ_1) and any fixed $\varepsilon > 0$, the following property:

(1.55) $$\lim_{\Delta t \to 0} \frac{1}{\Delta t} \int_{|x-y| > \varepsilon} p(\Delta t, x, y) dy = 0.$$

This condition is the same as

(1.55') $$\int_{|x-y| > \varepsilon} p(\Delta t, x, y) dy = o(\Delta t).$$

Now let

(1.56) $$\lim_{\Delta t \to 0} \frac{1}{\Delta t} \int_{|x-y| < \varepsilon} (y-x) p(\Delta t, x, y) dy = M_{\delta x},$$

(1.57) $$\lim_{\Delta t \to 0} \frac{1}{\Delta t} \int_{|x-y| < \varepsilon} (y-x)^2 p(\Delta t, x, y) dy = V_{\delta x}.$$

A markov process which satisfies conditions (1.55) and (1.55') is called a diffusion process. Let f(x) be an arbitrary function on J with a continuous second derivative on (γ_0, γ_1). Then

(1.58) $$\lim_{\Delta t \to 0} \frac{1}{\Delta t} \left[\int p(\Delta t, x, y) f(y) dy - f(x) \right] = \frac{V_{\delta x}}{2} \frac{d^2 f(x)}{dx^2} + M_{\delta x} \frac{df(x)}{dx}.$$

Proof: For a given x and an arbitrary $\delta > 0$, if we choose sufficiently small $\varepsilon > 0$, we have, for y such that $|y-x| < \varepsilon$,

$$f'(x)(y - x) + \{f''(x) - \delta\} \frac{(y-x)^2}{2}$$

$$\leq f(y) - f(x)$$

$$\leq f'(x)(y - x) + \{f''(x) + \delta\} \frac{(y-x)^2}{2}.$$

Note that

$$\int_{|y-x|<\varepsilon} p(\Delta t,\ x,\ y)f(y)dy - f(x)$$

$$= \int_{|y-x|<\varepsilon} \{f(y) - f(x)\}p(\Delta t,\ x,\ y)dy + o(\Delta t).$$

(The last term $o(\Delta t)$ is due to (1.55').) Therefore we have the following inequlities

$$f'(x) \int_{|y-x|<\varepsilon} (y-x)p(\Delta t,\ x,\ y)dy$$

$$+ \frac{f''(x)+\delta}{2} \int_{|y-x|<\varepsilon} (y-x)^2 p(\Delta t,\ x,\ y)dy + o(\Delta t)$$

$$\geqq \int_{|y-x|<\varepsilon} p(\Delta t,\ x,\ y)f(y)dy - f(x) \geqq$$

$$f'(x) \int_{|y-x|<\varepsilon} (y-x)p(\Delta t,\ x,\ y)dy$$

$$+ \frac{f''(x)-\delta}{2} \int_{|y-x|<\varepsilon} (y-x)^2 p(\Delta t,\ x,\ y)dy + o(\Delta t).$$

Dividing by Δt, taking the limit as $\Delta t \to 0$, and rearranging, we have

$$M_{\delta x}\ f''(x) + \frac{V_{\delta x}}{2}\ f''(x) + \frac{V_{\delta x}}{2}\ \delta$$

$$\geqq \lim_{\Delta t\to 0} \frac{1}{\Delta t} \left\{ \int f(y)p(\Delta t,\ x,\ y)dy - f(x) \right\}$$

$$\geqq M_{\delta x}f'(x) + \frac{V_{\delta x}}{2}\ f''(x) - \frac{V_{\delta x}}{2}\ \delta .$$

Since δ can be arbitrary small, the inequalities can be removed and we have (1.58).

Remark 1. In the drivation of equation (1.58), we did not need to assume that

$$\lim_{\Delta t\to 0} \frac{1}{\Delta t} \int (y-x)^k (\Delta t,\ x,\ y)dy = 0 \qquad \text{for } k > 2.$$

This property is in fact a consequence of condition (1.55).

Remark 2. We can weaken condition (1.55). It can be shown that (1.55) is a consequence of the continuity of the sample paths (trajectories). In other words, it is sufficient to assume that the trajectories are continuous, (see Ray, 1956).

Remark 3. In (1.56) and (1.57) we assumed that $V_{\delta x}$ and $M_{\delta x}$ are independent of time. A process of this kind is called "time homogeneous". However we can weaken this condition too. Now suppose that $p(\cdot, x, y)$ is not time homogeneous, and let $p(t_1, x; t_2, y)$ be the probability density that the value of X_t at $t = t_1$ is x and it is y at $t = t_2$, i.e.,

$$p(t_1, x; t_2, y) = \text{Prob}\{X_{t_2} = y \mid X_{t_1} = x\} .$$

Then instead of (1.54) and (1.55), we assume

(1.56')
$$\lim_{\Delta t \to 0} \frac{1}{\Delta t} \int_{|x-y|<\varepsilon} (y-x)p(t, x; t+\Delta t, y)dy = M_{\delta x}(t)$$

and

(1.57')
$$\lim_{\Delta t \to 0} \frac{1}{\Delta t} \int_{|x-y|<\varepsilon} (y-x)^2 p(t, x; t+\Delta t, y)dy = V_{\delta x}(t).$$

With these limits, the equation corresponding to (1.58) is

(1.58')
$$\lim_{\Delta t \to 0} \frac{1}{\Delta t} \int p(t, x; t+\Delta t, y)f(y)dy - f(x)$$

$$= \frac{V_{\delta x}(t)}{2} \frac{d^2 f(x)}{dx^2} + M_{\delta x}(t)\frac{df(x)}{dx} .$$

Remark 4. The formulae given in (1.56), (1.57), (1.56') and (1.57') may appear to be purely theoretical, but in many real situations, the two coefficients are given from their empirical meaning and the nature of the process.

For an arbitrary function f(x) on J, define

$$u(t, x) \doteq \int p(t, x, y)f(y)dy = T_t f(x).$$

This can be considered as a linear transformation on the set of all functions on J. Note that

$$T_0 f(x) = \int p(0, x, y)f(y)dy = \int \delta(x-y)f(y)dy = f(x)$$

where δ is Dirac's delta function which has the following property, for an arbitrary continuous function

$$\int_{-\infty}^{\infty} f(x)\delta(x-y)\,dx = f(y).$$

Now we can write equation (1.58) as

(1.59)
$$\lim_{\Delta t \to 0} \frac{1}{\Delta t}\left[T_{\Delta t} - I\right] = \frac{V_{\delta x}}{2}\frac{d^2}{dx^2} + M_{\delta x}\frac{d}{dx}$$

where I is the identity transformation. Let us denote by A the operator on the right side of (1.59). From the Chapman-Kolmogorov equation (1.53),

(1.60)
$$T_{t+s} = T_t T_s = T_s T_t.$$

Equation (1.59) together with (1.60) gives

(1.61)
$$\lim_{\Delta t \to 0} \frac{1}{\Delta t}\left[T_{t+\Delta t} - T_t\right] = \lim_{\Delta t \to 0}\frac{1}{\Delta t}\left[T_{\Delta t} - I\right]T_t = AT_t$$

$$= \lim_{\Delta t \to 0}\frac{1}{\Delta t}T_t\left[T_{\Delta t} - I\right] = T_t A.$$

Note that the operators T_t and A commute. Therefore we have

(1.62)
$$\frac{dT_t f(x)}{dt} = AT_t f(x) = T_t A f(x).$$

The left and the middle in the above equation can be rewritten as

(1.63)
$$\frac{\partial u(t,x)}{\partial t} = Au(t,x) = \left\{\frac{V_{\delta x}}{2}\frac{\partial^2}{\partial x^2} + M_{\delta x}\frac{\partial}{\partial x}\right\}u(t,x)$$

where $u(t,x) = \int p(t,x,y)f(y)\,dy$. This is called the Kolmogorov backward equation, (hereafter denoted by KBE).

How do we solve this equation? By analogy to equation (1.6) and (1.7), a formal solution would be

$$u(t,x) = e^{tA}u(0,x)$$

and since $u(0,x) = f(x)$, this becomes

(1.64)
$$u(t,x) = e^{tA}f(x).$$

Again, in analogy to the power series expansion (1.8), we have

(1.65) $\quad e^{tA} = I + tA + \dfrac{t^2}{2!} A^2 + \dfrac{t^3}{3!} A^3 + \cdots + \dfrac{t^k}{k!} A^k + \cdots$

and hence

(1.66) $\quad u(t, x) = \left\{ I + tA + \dfrac{t^2}{2!} A^2 + \cdots + \dfrac{t^k}{k!} A^k + \cdots \right\} f(x)$

where

$$A^k = \underbrace{\left\{ \dfrac{V_{\delta x}}{2} \dfrac{d^2}{dx^2} + M_{\delta x\frac{d}{dx}} \right\} \left\{ \dfrac{V_{\delta x}}{2} \dfrac{d^2}{dx^2} + M_{\delta x dx} \right\} \cdots \left\{ \dfrac{V_{\delta x}}{2} \dfrac{d^2}{dx^2} + M_{\delta x dx} \right\}.}_{k \text{ times}}$$

Suppose each of the function $\{e_k(x), \ k = 0, 1, 2, \cdots\}$ satisfies

$$Ae_k(x) + \lambda_k e_k(x) = 0$$

where λ_k is a constant. These sets of functions and constants are called eigenfunctions and eigenvalues of operator A. Use of these sets is analogous to that of linear algebra. If we can express an arbitrary function $f(x)$ in terms of $\{e_k(x)\}$, i.e.,

(1.67) $\qquad\qquad f(x) = \sum\limits_{k} a_k e_k(x)$

where a_k are constants, (a Fourier series), and applying the operator e^{tA} to (1.67), we have

(1.68) $\quad e^{At} f(x) = \sum\limits_{k} a_k e_k(x) + t \sum a_k A e_k(x)$

$$+ \dfrac{t^2}{2!} \sum a_k A^2 e_k(x) + \dfrac{t^3}{3!} \sum a_k A^3 e_k(x) + \cdots .$$

Since

$$A^n e_k(x) = A^{n-1} (-\lambda_k) e_k(x) = \cdots = (-\lambda_k)^n e_k(x) ,$$

(1.68) becomes

$$e^{At} f(x) = \sum\limits_{k=0}^{\infty} a_k e^{-\lambda_k t} e_k(x).$$

Therefore we have

(1.69) $\qquad\qquad u(t, x) = \sum\limits_{k=0}^{\infty} a_k e^{-\lambda_k t} e_k(x).$

More concrete cases of this solution will be discussed later.

1.6 Dynkin's formula

We have derived the KBE from the transition probability function $p(t, x, y)$. We considered a small time interval $(0, \Delta t)$ and investigated the behaviors of all paths at <u>fixed</u> <u>times</u> 0 and Δt. There is an alternative approach. We look at the behavior of a path when it leaves a <u>fixed</u> <u>area</u>, instead of a fixed time.

Let U be an open set in thé state space, and let $\tau_\omega^U(x)$ be the first "exit" time of path ω from U, given that ω starts from $x \in U$. Then it can be shown that

$$(1.70) \qquad \lim_{U \downarrow x} \frac{E_x \left\{ f(x_{\tau_\omega^U(x)}, \omega) \right\} - f(x)}{E\{\tau_\omega^U(x)\}} = Af(x)$$

where $A = \dfrac{V_{\delta x}}{2} \dfrac{\partial^2}{\partial x^2} + M_{\delta x} \dfrac{\partial}{\partial x}$, $E_x\{\cdot\}$ is the expectation for sample paths starting from x, and $x_{\tau_\omega^U(x)}$ indicates the location of path ω at the first exit moment. See Dynkin (1965, chapt. V).

CHAPTER 2

POPULATION GENETICS MODELS

Consider a population consisting of 2N genes at a single locus
(N diploid individuals) which are either A or a, where N is fixed
number.

Let us first consider a simple case of pure random drift, hence
ignoring mutation pressure, selection force, etc.

2.1 Wright's model

At each generation 2N genes are drawn randomly from the gene pool
with replacement. If the parent population consists of j A-genes and
(2N-j)a-genes, each drawing results in A or a with probabilities j/2N
of (2N-j)/2N respectively. In other words, the process is a Markov
chain with binomial transition probabilities

(2.1)
$$P_{ij} = \binom{2N}{i}\left(\frac{j}{2N}\right)^i \left(1 - \frac{j}{2N}\right)^{2N-i}$$

where $\binom{2N}{i}$ is the binomial coefficient, $\frac{2N!}{i!(2N-i)!}$. See Wright (1931).

Let X_t be the frequency of A-genes at time t. The state space of
X_t is [0, 1/2N, 2/2N, 3/2N, $\cdots$, (2N-1)/2N, 1]. For the time being,
assume that the trajectories of X_t can be approximated by paths which
are continuous. Then we have a diffusion process. (see Condition
(1.55) and Remark 2.)

To obtain the explicit form of the KBE for the above Wright model,
we need to calculate $M_{\delta x}$ and $V_{\delta x}$. Suppose that the (discrete) generations change in the time interval Δt with probability $\Delta t + o(\Delta t)$. If
the generations change, the variance of the gene frequency change is

$$(2.2) \quad \sum_{i=0}^{2N} \left(\frac{j}{2N}\right)^2 \binom{2N}{i} x^i (1-x)^{2N-i} - \left\{ \sum_{i=0}^{2N} \left(\frac{i}{2N}\right)\binom{2N}{i} x^i (1-x)^{2N-i} \right\}^2$$

$$= \frac{x(1-x)}{2N}$$

where x is the value of X_t immediately prior to the generation change. Obviously the mean change is zero. Therefore

$$(2.3) \quad V_{\delta x} = \lim_{\Delta t \to 0} \frac{1}{\Delta t} E\left\{ (X_{t+\Delta t} - X_t)^2 \right\} = \frac{x(1-x)}{2N}$$

and

$$(2.4) \quad M_{\delta x} = \lim_{\Delta t \to 0} \frac{1}{\Delta t} E\left\{ (X_{t+\Delta t} - X_t) \right\} = 0$$

where $E\{\ \}$ is the expectation. The KBE for the pure random drift case is therefore

$$(2.5) \quad \frac{\partial u(t, x)}{\partial t} = \frac{x(1-x)}{4N} \frac{\partial^2 u(t, x)}{\partial x^2}$$

where

$$u(t, x) = \int_0^1 p(t, x, y) f(y) dy$$

with an arbitrary function $f(x)$.

2.2 Feller's model

There are alternative ways of obtaining KBE (2.5). W. Feller's idea is as follows: We choose the time scale so that the time required for one generation is

$$\Delta t = 1/N .$$

Then

$$(2.6) \quad \lim_{\substack{\Delta t \to 0 \\ (N \to \infty)}} \frac{1}{\Delta t} E\left\{ (X_{t+\Delta t} - X_t)^2 \right\} = \lim_{\substack{\Delta t \to 0 \\ (N \to \infty)}} \frac{1}{\Delta t} \frac{x(1-x)}{2N}$$

$$(\Delta t = 1/N)$$

$$= \frac{x(1-x)}{2} = V_{\delta x} .$$

Obviously $M_{\delta x} = 0$. Therefore the KBE is

$$(2.7) \quad \frac{\partial u(t, x)}{\partial t} = \frac{x(1-x)}{4} \frac{\partial^2 u(t, x)}{\partial x^2} .$$

The new time interval of duration t then corresponds to N = t/Δt generations, (see Feller, 1951).

If we make the time substitution in (2.7) by letting t' = Nt, KBE (2.7) becomes identical to KBE (2.5). In Feller's derivation, we can also show that for a fixed j/2N = x and a fixed δ > 0

(2.8)
$$\sum_{|\frac{i}{N} - x| > \delta} P_{ij} = \sum_{|\frac{i}{N} - x| > \delta} \binom{2N}{i}\left(\frac{j}{2N}\right)^i \left(1 - \frac{j}{2N}\right)^{2N-j}$$

$$= o\left(\frac{1}{N}\right) .$$

This condition is equivalent to condition (1.55).

2.3 Moran's model

There is still another way given by Moran (1958). Suppose that each existing gene is subject to death with equal probability Δt+o(Δt) in time Δt. Suppose furthermore that when a death occurs it is immediately replaced by a copy (newly born) of a randomly chosen gene among the population immediately before the death. The probability that the value of X_t is increased by 1/2N (or decreased by 1/2N) in Δt is

$$2N\Delta t x (1 - x) + o(\Delta t),$$

and therefore

$$\lim_{\Delta t \to 0} \frac{1}{\Delta t} E\left\{(X_{t+\Delta t} - X_t)^2\right\} = \lim_{\Delta t \to 0} \frac{1}{\Delta t}[\frac{1}{2N}]^2 [2N\Delta t x (1 - x) + o(\Delta t)]$$

$$= \frac{x(1 - x)}{N} = V_{\delta x} .$$

The KBE for the Moran model would be

(2.9)
$$\frac{\partial u(t, x)}{\partial t} = \frac{x(1 - x)}{2N} \frac{\partial^2 u(t, x)}{\partial x^2} .$$

Problem 1. Note that the right sides of (2.5) and (2.9) differ by a factor of 2, i.e., $V_{\delta x} = x(1-x)/4N$ in (2.5) while $V_{\delta x} = x(1-x)/2N$ in (2.9). Why?

2.4 Variable population size

In the above genetic models, it is assumed that the population size is constant. But we can weaken this restriction on the assumption that the trajectories of the process X_t are still approximately continuous. In the following three types of modification, we shall assume that the population size is independent of the genetic constitution of the population. Therefore the size is controlled by other factors.

First suppose that the population size changes deterministically according to a preassigned rule. If we denote the size at time t by N_t, then using formula (1.57'), we have

(2.10)
$$V_{\delta x}(t) = \frac{x(1 - x)}{2N_t} \quad .$$

Now suppose that the population size changes according to a certain probability law, instead of a deterministic rule. Let $p(N)$ be the distribution from which the population size is drawn at each generation independently. Then the expected variance of the gene frequency change would be

(2.11)
$$V_{\delta x} = \int \frac{x(1 - x)}{2N} \, dp(N) = \frac{x(1 - x)}{2} \int \frac{1}{N} \, dp(N)$$

$$= \frac{x(1 - x)}{2\hat{N}}$$

where $\hat{N} = \dfrac{1}{\int \frac{1}{N} \, dp(N)}$. In equation (2.11), $\int dp(N) = \int p'(N) \, dN$ if

continuous, $= \Sigma \, p(N)$ if discrete. Next assume that the size distribution also changes in time, and denote it by $p(t, N_t)$. Then

(2.12)
$$V_{\delta x}(t) = \int \frac{x(1 - x)}{2N_t} \, dp(t, N_t) = \frac{x(1 - x)}{2} \int \frac{1}{N_t} \, dp(t, N_t)$$

$$= \frac{x(1 - x)}{2\hat{N}_t}$$

where $\hat{N}_t = 1 \Big/ \!\!\int \frac{1}{N_t} \, dp(t, N_t)$.

Therefore, the KBEs for (2.10),(2.11) and (2.12) are respectively

(2.13)
$$\frac{\partial u(t, x)}{\partial t} = \frac{x(1 - x)}{4N_t} \frac{\partial^2 u(t, x)}{\partial x^2} \, ,$$

(2.14)
$$\frac{\partial u(t, x)}{\partial t} = \frac{x(1 - x)}{4\hat{N}} \frac{\partial^2 u(t, x)}{\partial x^2} \, ,$$

(2.15)
$$\frac{\partial u(t, x)}{\partial t} = \frac{x(1 - x)}{4\hat{N}_t} \frac{\partial^2 u(t, x)}{\partial x^2}$$

For all these cases, the time unit can be changed so that they have an identical form of the KBE. Namely, the time substitution to be taken for (2.13) is

(2.16)
$$\tau = \int_0^t \frac{1}{2N_\xi} \, d\xi \; ;$$

for (2.14)

(2.17)
$$\tau = \frac{t}{2\hat{N}_\xi} \; ;$$

for (2.15)

(2.18)
$$\tau = \int_0^t \frac{1}{2\hat{N}_\xi} \, d\xi \; .$$

With these time changes, the KBE becomes

(2.19)
$$\frac{\partial u(\tau, x)}{\partial \tau} = \frac{x(1 - x)}{2} \frac{\partial^2 u(\tau, x)}{\partial x^2} \; .$$

The significance of this result is as follows: The time measured by the τ defined in (2.16) $\sim$ (2.18) transforms the processes into an identical one. Therefore the trajectories for (2.13), (2.14), (2.15) and (2.19) are the same, and when we look at them with time τ, they are the same process.

In the these genetic models, we assumed that no deterministic pressure operates. As a natural step towards more general models, we may incorporate mutation pressure.

2.5 Wright's model with mutation

In addition to the assumptions imposed in the model stated in equation (2.1), we assume that A-genes mutate to a-genes with probability uΔt in the time interval Δt, and a-genes to A-genes with probability vΔt.

$$A \underset{v}{\overset{u}{\rightleftarrows}} a$$

Fig. 2.1

Suppose that the random variable X_t = frequency of A has value x at some generation t. Then if the generation changes in (t, t+Δt), the expected value of $X_{t+\Delta t}$ is

$$(1 - u\Delta t)x + v\Delta t(1 - x) + o(\Delta t)$$

and therefore the mean change of the value of X_t is

$$M_{\delta x} = \lim_{\Delta t \to 0} \frac{1}{\Delta t} \{(1 - u\Delta t)x + v\Delta t(1 - x) + o(\Delta t) - x\} = v - (u + v)x.$$

Since the second order effect of mutation and random sampling of gametes vanishes in the limit, the variance is

$$V_{\delta x} = \frac{x(1 - x)}{2N} .$$

Hence the KBE is

$$\frac{\partial u(t, x)}{\partial t} = \frac{x(1 - x)}{4N} \frac{\partial^2 u(t, x)}{\partial x^2} + \{v - (u + v)x\}\frac{\partial u(t, x)}{\partial x}$$

where

$$u(t, x) = \int p(t, x, y)f(y)dy.$$

2.6 A model of irreversible mutation or a model of infinite alleles

In the above model, if we let $v = 0$, the A-gene mutates to a, but not in the reverse direction. The KBE is then

$$(2.20) \qquad \frac{\partial u(t, x)}{\partial t} = \frac{x(1 - x)}{4N} \frac{\partial^2 u(t, x)}{\partial x^2} - ux \frac{\partial u(t, x)}{\partial x} .$$

This model can be interpreted alternatively as follows: The number of possible states to mutate to is very large, and therefore it is nearly impossible to have a mutation which will give rise to a preexisting type. A particular allele under consideration will be designated by A and the rest of the alleles by a collectively. This model has been studied in detail by Kimura and Crow (1964), and is often referred to as the Kimura-Crow infinite allele model.

2.7 A selection model

Disregarding mutation, now suppose that A- and a-genes are selectively different and that the relative fitnesses are 1+s and 1 respectively. If the frequency of A-genes is x at some generation, and if the effects of genes are additive in diploids, the expected frequency at the next can calculated as follows:

Gene	A	a
Frequency	x	1 - x
Fitness	1 + s	1

$$(1 + s)x \quad 1 - x$$

Average fitness $= \bar{W} = (1 + s)x + (1 - x) = 1 + sx$

and

A-frequency in next generation $= x' = \dfrac{(1 + s)x}{1 + sx}$.

Hence the mean change is

$$M_{\delta x} = x' - x = \frac{(1 + s)x}{1 + sx} - x = \frac{s}{1 + sx} x(1 - x)$$

and the variance is $V_{\delta x} = x(1-x)/2N$. The KBE is

$$(2.21) \qquad \frac{\partial u}{\partial t} = \frac{x(1 - x)}{4N} \frac{\partial^2 u}{\partial x^2} + \frac{s}{1 + sx} x(1 - x)\frac{\partial u}{\partial x} .$$

If $|s| \ll 1$, $s/(1+sx) \approx s$ and

$$(2.22) \qquad \frac{\partial u}{\partial t} = \frac{x(1 - x)}{4N} \frac{\partial^2 u}{\partial x^2} + sx(1 - x)\frac{\partial u}{\partial x} .$$

Remark 1. One may wonder at the significance of the difference between KBE (2.21) and (2.22). It seems important to realize that both KBE (2.21) and (2.22) correspond to a bona fide diffusion process. Therefore it is incorrect to say, for example, that KBE (2.22) gives a better approximation (or worse) than (2.21). The accuracy depends on the models. The KBE (2.21) will tend to give a better result if the generations are discrete, while the KBE (2.22) gives a better re-sult for a time continuous model and for some models of competitive selection (Mather (1969), and Nei and Yokoyama (1976)).

Remark 2. As implied in the above remark, for each differential operator, say,

$$A(t, x) = a(t, x)\frac{\partial^2}{\partial x^2} + b(t, x)\frac{\partial}{\partial x}$$

there exists at least one diffusion process whose KBE is

$$\frac{\partial u}{\partial t} = A(t, x)u .$$

This is a converse of (1.62) and (1.63), in the sense that the opera-tor A in (1.62) and (1.63) is derived from a given diffusion process, while here we derive the process from a given differential operator A of second order. This is the so-called Hille-Yoshida semi-group theory. The construction of a process from a given operator Λ is given by

$$p(t, x, y) = e^{tA} = I + tA + \frac{t^2}{2!} A^2 + \frac{t^3}{3!} A^3 + \cdots$$

where p(t, x, y) is the transition probability in the constructed process, (see Feller, 1965, chapt. X).

2.8 A model of dominance

Let

Genotype	AA	Aa	aa
Frequency	x^2	$2x(1 - x)$	$(1 - x)^2$
Fitness	$1 + s$	$1 + sh$	1

where h is a measure of the degree of dominance. Then the average fitness is

$$\bar{w} = (1 + s)x^2 + (1 + sh)2x(1 - x) + (1 - x)^2$$

$$= 1 + sx^2 + 2shx(1 - x)$$

and since

$$x' = \frac{(1 + s)x^2 + (1 + sh)x(1 - x)}{\bar{w}} \quad,$$

we have

$$M_{\delta x} = x' - x = sx(1 - x)\frac{\{x + 2h(1 - x)\}}{\bar{w}}$$

so that the KBE is

$$(2.23) \qquad \frac{\partial u}{\partial t} = \frac{x(1 - x)}{4N} \frac{\partial^2 u}{\partial x^2} + sx(1 - x)\frac{\{x + 2h(1 - x)\}}{\bar{w}} \frac{\partial u}{\partial x} .$$

2.9 Birth-and-death processes

We assume that each member independently gives rise to a progeny with probability $\theta \Delta t$ and it is also subject to death with probability $\delta \Delta t$ in Δt. Let X_t be the population number and assume that the process X_t can be approximated by a continuous state space process. Then it is a diffusion type and it has

$$V_{\delta x} = \lim_{\Delta t \to 0} \frac{1}{\Delta t} [\theta \Delta t + \delta \Delta t]x = (\theta + \delta)x$$

and

$$M_{\delta x} = \lim_{\Delta t \to 0} \frac{1}{\Delta t} [\theta \Delta t - \delta \Delta t] x = (\theta - \delta) x .$$

The KBE is

(2.24) $$\frac{\partial u}{\partial t} = \frac{1}{2}(\theta + \delta) x \frac{\partial^2 u}{\partial x^2} + (\theta - \delta) x \frac{\partial u}{\partial x} .$$

More generally, if the mean change and variance of the above process depend on time, the KBE is

(2.25) $$\frac{\partial u}{\partial t} = \frac{1}{2}[\theta(t) + \delta(t)] x \frac{\partial^2 u}{\partial x^2} + [\theta(t) - \delta(t)] x \frac{\partial}{\partial x} .$$

2.10 Density or frequency dependent process

We still assume the independent reproduction of individuals, and furthermore that the probability of giving one birth is $\theta(x) \Delta t$ and the probability of death is $\delta(x) \Delta t$. Note that this process differs from the previous one in having birth and death rates $\theta(x)$ and $\delta(x)$ which are dependent on x, the population number. The KBE is

(2.26) $$\frac{\partial u}{\partial t} = \frac{1}{2}[\theta(x) + \delta(x)] x \frac{\partial^2 u}{\partial x^2} + [\theta(x) - \delta(x)] x \frac{\partial u}{\partial x} .$$

For example assume that the birth and death probabilities are respectively $(a-bx) \Delta t$ and $(c+dx) \Delta t$, where a, b, c and d are positive constants. Then

$$M_{\delta x} = \{(a - c) - (d + b)x\}x,$$

$$V_{\delta x} = \{(a + c) + (d - b)x\}x$$

and

$$\frac{\partial u}{\partial t} = \{(a + c) + (d - b)x\}x \frac{\partial^2 u}{\partial x^2} + \{(a - c) - (d + b)x\}x \frac{\partial u}{\partial x} .$$

Remark 3. Note that the coefficients $M_{\delta x}$ and $V_{\delta x}$ or $M_{\delta x}(t)$ and $V_{\delta x}(t)$ of the KBE's are not required to be differentiable. However, in the case of the Kolmogorov forward equations, the differentiability of the coefficients is required. Furthermore the coefficients of a KBE can have any finite number of discountinuities. For example, consider the following scheme.

Frequency dependent selection

Gene	A	a	
Frequency	x	1 - x	
Fitness	1 + s	1 - s	if $x < \frac{1}{2}$
	1 - s	1 + s	if $x > \frac{1}{2}$

Then

$$M_{\delta x} = \frac{2sx(1 - x)}{1 + s(2x - 1)} \qquad \text{if } x < \frac{1}{2} ,$$

$$M_{\delta x} = \frac{-2sx(1 - x)}{1 + s(1 - 2x)} \qquad \text{if } x > \frac{1}{2}$$

and

$$V_{\delta x} = \frac{x(1 - x)}{2N} .$$

Hence the KBE is

$$\frac{\partial u}{\partial t} = \frac{x(1 - x)}{4N} \frac{\partial^2 u}{\partial x^2} + \frac{2sx(1 - x)}{1 + s(2x - 1)} \frac{\partial u}{\partial x} \qquad \text{for } x < \frac{1}{2} ,$$

$$= \frac{x(1 - x)}{4N} \frac{\partial^2 u}{\partial x^2} - \frac{2sx(1 - x)}{1 + s(1 - 2x)} \frac{\partial u}{\partial x} \qquad \text{for } x > \frac{1}{2} .$$

2.11 Time inhomogeneous process

Let $p(t_1, x; t_2, y)$ with $t_1 < t_2$ be the probability density that the value of the random variable X_t is y at time t_2, given that it is x at time t_1. And let

$$T(t_2, t_1)f(x) = u(t_1, t_2, x) = \int p(t_1, x; t_2, y)f(y)dy$$

for an arbitrary $f(x)$. First note that

$$p(t_1, x; t_2, y) = \int p(t_1, x; s, \xi)p(s, \xi; t_2, y)d\xi .$$

Hence

$$\int p(t_1, x; t_2, y)f(y)dy$$

$$= \int \{\int p(t_1, x; s, \xi)p(s, \xi; t_2, y)d\xi\}f(y)dy$$

$$= \int p(t_1, x; s, \xi)\{\int p(s, \xi; t_2, y)f(y)dy\}d\xi$$

$$= \int p(t_1, x; s, \xi)\{T(s, t_2)f(\xi)\}d\xi$$

$$= T(t_1, s)\{T(s, t_2)f(x)\}$$

$$= T(t_1, s)T(s, t_2)f(x).$$

Therefore

$$T(t_2, t_1) = T(t_2, s)T(s, t_1) \qquad \text{for } t_1 < s < t_2.$$

Then

$$\frac{1}{\Delta t}\{T(t_2+\Delta t, t_1) - T(t_2, t_1)\}$$

$$= \frac{1}{\Delta t}\{T(t_2+\Delta t, t_2)T(t_2, t_1) - T(t_2, t_1)\}$$

$$= \frac{1}{\Delta t}\{T(t_2+\Delta t, t_2) - I\}T(t_2, t_1).$$

Therefore letting Δt become small, we have

$$\frac{\partial T(t_2, t_1)}{\partial t_2} = A(t_2)T(t_2, t_1)$$

in which

$$A(t_2) = \frac{V_{\delta x}(t_2)}{2}\frac{\partial^2}{\partial x^2} + M_{\delta x}(t_2)\frac{\partial}{\partial x}$$

where $M_{\delta x}(t)$ and $V_{\delta x}(t)$ are given by formulae (1.54') and (1.55'). If we let $t_1 = 0$ and

$$u(t, x) = u(0, t, x) = \int p(0, x; t, y)f(y)dy$$

the KBE is

$$\frac{\partial u}{\partial t} = \frac{V_{\delta x}(t)}{2}\frac{\partial^2 u}{\partial x^2} + M_{\delta x}(t)\frac{\partial u}{\partial x}$$

where $V_{\delta x}(t)$ and $M_{\delta x}(t)$ are now functions of time.

2.12 A model of random environment

Assume that the relative fitnesses of the A- and a- alleles change in generations.

Gene	A	a
Fitness	$1 + s_t/2$	$1 - s_t/2$

where $E\{s_t\} = \bar{s}$ and $\text{Var}\{s_t\} = E\{s_t^2\} - E\{s_t\}^2 = \sigma^2$. We assume that the s_t changes only in different generations, but it remains constant for each generation. In other words, at each generation, the value of s_t is drawn from a random variable of mean $\bar{s}$ and variance σ^2. As before we assume that the sample paths are continuous, and the generations are discrete and they change in time Δt with probability $\Delta t + o(\Delta t)$. Then

$$\Delta x = \frac{s_t x(1 - x)}{1 - s_t (\frac{1}{2} - x)} = x(1 - x)\left\{ s_t + s_t^2 (\frac{1}{2} - x) + s_t^3 (\frac{1}{2} - x)^2 + \cdots \right\}.$$

Therefore, if we ignore the higher order terms in s_t,

$$E(\Delta x) = x(1 - x)\left\{ \bar{s} + (\sigma^2 + \bar{s}^2)(\frac{1}{2} - x) \right\}.$$

We assume that the population is infinitely large so that the gene frequency change is caused only by the selection due to the random fluctuation of the environment. Then the variance is

$$V_{\delta x} = E(\Delta x^2) = x^2(1 - x)^2 E\left\{ \frac{s_t^2}{(1 - \frac{s_t}{2} + s_t x)^2} \right\},$$

and if $|s_t|$ is sufficiently small

$$V_{\delta x} \approx (\sigma^2 + \bar{s}^2)x^2(1 - x)^2.$$

The KBE is

$$(2.23) \qquad \frac{\partial u}{\partial t} = \frac{(\sigma^2 + \bar{s}^2)x^2(1 - x)^2}{2} \frac{\partial^2 u}{\partial x^2}$$

$$+ x(1 - x)\left\{ \bar{s} + (\sigma^2 + \bar{s}^2)(\frac{1}{2} - x) \right\}\frac{\partial u}{\partial x}.$$

CHAPTER 3

CLASSIFICATION OF BOUNDARIES

Let (r_0, r_1) be an open interval which is the whole or a part of the state space where the process is taking place. For example:

Fig. 3.1 Illustrating an open interval
which is a part or the entire space.

Our aim here is to know the behavior of the trajectories (sample paths) in a neighborhood of r_0 or r_1. According to Feller (1952), there are four kinds of behavior at a boundary point, i.e., r_0 or r_1. To express the criteria for them we need to introduce the function

(3.1)
$$w(x) = e^{-\int_{x_0}^{x} \frac{2M_{\delta\xi}}{V_{\delta\xi}} d\xi}$$

where $x_0 \in (r_0, r_1)$ and fixed, and $M_{\delta\xi}$ and $V_{\delta\xi}$ are the coefficients of a KBE under consideration.

3.1 Regular boundary

The boundary $r_j (j = 0, 1)$ is called "REGULAR" if for any $x_0 \in (r_0, r_1)$,

(3.2)
$$w(x) \in L(x_0, r_j) \qquad \text{and}$$

(3.3)
$$\frac{1}{V_{\delta x} w(x)} \in L(x_0, r_j)$$

where $f(x) \in L(a, b)$ means

$$-\infty < \int_a^b f(x) dx < \infty \qquad \text{(integrable).}$$

In this case, a trajectory can reach r_j from the interior of (r_0, r_1) and can return to the interior within a finite length of time.

3.2 Exit boundary

The boundary r_j is an "EXIT" boundary, if

(3.4)
$$\frac{1}{V_{\delta x} w(x)} \notin L(x_0, r_j) \qquad \text{and}$$

(3.5)
$$w(x) \int_{x_0}^x \frac{d\xi}{V_{\delta \xi} w(\xi)} \in L(x_0, r_j).$$

In this case, a trajectory can reach r_j from the interior within a finite time, but it can not return to the interior.

3.3 Entrance boundary

The r_j is an "ENTRANCE" boundary if r_j is not regular but

(3.6)
$$\frac{1}{V_{\delta x} w(x)} \in L(x_0, r_j) \qquad \text{and}$$

(3.7)
$$\frac{1}{V_{\delta x} w(x)} \int_{x_0}^x w(\xi) d\xi \in L(x_0, r_j).$$

A trajectory can enter from r_j to the interior, but it can not return from the interior to r_j.

3.4 Natural boundary

The r_j is a "NATURAL" boundary in all other cases. A trajectory can neither reach from the interior to r_j, nor from r_j to the interior. It is called a "TRAP" point.

3.5 Nature of boundary

The significance of these boundaries are:
(i) If r_j is a regular or exit boundary, a trajectory has a possibility to reach it from the interior within a finite time. (For this reason,

the regular and exit boundaries are called "ACCESSIBLE".)

(ii) If r_j is an entrance or natural boundary, a trajectory will never reach it, and thus the natural and entrance boundaries are called "INACCESSIBLE".

(iii) If both r_0 and r_1 are inaccessible, there exists one and only one process in (r_0, r_1) obeying a given KBE.

(iv) Whenever both r_0 and r_1 are accessible, there exists an "absorbing barrier process" in (r_0, r_1) obeying a given KBE. By absorbing we mean that at the first <u>arrival</u> at a boundary, (i.e., as soon as it arrives at r_0 or r_1) it is terminated.

(v) Suppose that (ρ_0, ρ_1) is a proper subinterval of (r_0, r_1) where r_0 and r_1 are inaccessible and ρ_0 and ρ_1 are regular boundaries. We can construct a solution for the absorbing barrier process in (ρ_0, ρ_1).

Then as we let $\rho_0 \to r_0$ and $\rho_1 \to r_1$, the solution corresponding to the absorbing process converges to the unique solution described in (iii).

(vi) If a boundary $r_j (j = 0, 1)$ is accessible, we can construct a return process of the following nature. When a trajectory reaches the boundary r_j, it waits there for a time t with probability e^{-t/λ_j}. Then a jump occurs to point x in the interior or to the boundary r_k (k = 0, 1) according to the rule:

Probability of a jump to $r_k = P_{jk}$,

Probability of a jump to (x-dx/2, x+dx/2) with x ϵ (r_0, r_1)

$$= \tau_j p_j(x) dx$$

where $\displaystyle\int_{r_0}^{r_1} p_j(x) dx = 1$ and so τ_j = probability of a jump into (r_0, r_1) from r_j. Then

(3.8) $P_{j0} + P_{j1} + \tau_j \le 1$,

where the difference of the two sides in (3.8) accounts for the possibility of the process terminating at r_j.

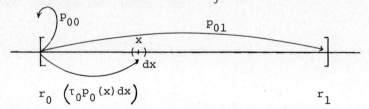

Fig. 3.2 Illustration of a return process, from r_0 to various locations.

After the jump, the trajectory starts from scratch. The solution of the KBE corresponding to this process is characterized by the two conditions

(3.9)
$$(1 - p_{00})u(t, r_0) - p_{01}u(t, r_1)$$

$$= \tau_0 \int_{r_0}^{r_1} u(t, y)p_0(y)\,dy - \lambda_0 Au(t, r_0)$$

and

(3.10)
$$- p_{10}u(t, r_0) + (1 - p_{11})u(t, r_1)$$

$$= \tau_1 \int_{r_0}^{r_1} u(t, y)p_1(y)\,dy - \lambda_1 Au(t, r_1)$$

where

$$A = \frac{V_{\delta x}}{2}\frac{d^2}{dx^2} + M_{\delta x}\frac{d}{dx} \; .$$

If $\lambda_0 = \lambda_1 = 0$ the process takes place in the open interval (r_0, r_1).
If both r_0 and r_1 are inaccessible, then no return process is possible. If r_0 is inaccessible, but r_1 is accessible, the above construction applies to the boundary r_1.

A special case is that

$$p_{00} + p_{01} + \tau_0 = 0$$

which implies

(3.11)
$$u(t, r_0) = 0.$$

In this case all trajectories terminate as soon as they reach the boundary r_0. A similar situation can apply to the boundary r_1. (See Feller (1954).)

3.6 Examples

Consider the pure random drift case whose KBE is given by (2.5), and let $r_0 = 0$ and $r_1 = 1$, where $(0, 1)$ is the whole space. Here

$$w(x) = e^{-2\int \frac{M_{\delta\xi}}{V_{\delta\xi}}} = e^0 = 1.$$

Obviously

$$w(x) \in L(0, \tfrac{1}{2}) \qquad \text{but}$$

$$\frac{1}{V_{\delta x}w(x)} = \frac{2N}{x(1-x)} \notin L(0, \tfrac{1}{2}) .$$

Hence $r_0 = 0$ is not regular.

$$\int_x \frac{d\xi}{V_{\delta\xi}w(\xi)} = \int_x \frac{2N}{\xi(1-\xi)} = 2N[\log(1-x) - \log x] .$$

Thus

$$w(x) \int_x \frac{d\xi}{V_{\delta\xi}w(\xi)} = 2N[\log(1-x) - \log x] .$$

The right side of the above formula is integrable in $(0, \tfrac{1}{2})$, therefore

$$\frac{1}{V_{\delta x}w(x)} \in L(0, \ \varepsilon) ,$$

$$w(x) \int_x \frac{d\xi}{V_{\delta\xi}w(\xi)} \in L(0, \ \varepsilon)$$

and by (3.4) and (3.5), $r_0 = 0$ is an exit boundary. Similarly we can show that $r_1 = 1$ is also an exit. This result is intuitively clear, because without mutation once a trajectory reaches a boundary it will never return to the interior of $(0, 1)$.

Next consider the case of irreversible mutation (or equivalently the infinite allele) model. The KBE is given by equation (2.20):

$$\frac{\partial u(t, x)}{\partial t} = \frac{x(1-x)}{4N} \frac{\partial^2 u(t, x)}{\partial x^2} - ux \frac{\partial u(t, x)}{\partial x} .$$

Here

$$w(x) = \exp \int \frac{4Nux}{x(1-x)} \, dx = \exp\{-4Nu \log(1-x)\}$$

$$= (1 - x)^{-4Nu} .$$

and so

$$\int w(x) \, dx = \frac{(1-x)^{1-4Nu}}{1 - 4Nu} .$$

Hence

(3.12) $w(x) \in L(0, \tfrac{1}{2})$ for all $4Nu$,

(3.13) $w(x) \in L(\tfrac{1}{2}, 1)$ if $4Nu < 1$,

(3.14) $\notin L(\tfrac{1}{2}, 1)$ if $4Nu > 1$,

$$\frac{1}{V_{\delta x}w(x)} = \frac{4N}{x(1-x)^{1-4Nu}}$$

and thus

(3.15)
$$\frac{1}{V_{\delta x}w(x)} \notin L(0, \tfrac{1}{2}) \qquad \text{for all } 4Nu,$$

(3.16)
$$\in L(\tfrac{1}{2}, 1) \qquad \text{for all } 4Nu.$$

By (3.13) and (3.16), $r_1 = 1$ is regular if $4Nu < 1$.

$$\frac{1}{V_{\delta x}w(x)} \int w(\xi)d\xi = \frac{4N}{x(1-x)^{1-4Nu}} \frac{(1-x)^{1-4Nu}}{(1-4Nu)}$$

(3.17)
$$= \frac{4N}{(1-4Nu)x} \in L(\tfrac{1}{2}, 1).$$

By (3.14), 3.16) and (3.17), $r_1 = 1$ is entrance, if $4Nu > 1$.

Therefore if $4Nu$ is less than 1, the boundary $x = 1$ is accessible and regular, and thus a path can reach the boundary within a finite time interval. This means that a temporary fixation of an allele in the whole population can occur, but the fixation is not permanent and the population will again become polymorphic unless the process is stopped. On the other hand, if $4Nu$ is greater than 1, the boundary $x = 1$ is not accessible and it is an entrance. In this case a population will never be fixed for a single allele, but is always polymorphic.

We will next examine the nature of the other boundary $x = 0$. Note that from (3.15) the integrability of

$$\frac{1}{V_{\delta x}w(x)} = \frac{4N}{x(1-x)^{1-4Nu}}$$

near $x = 0$ depends only on $1/x$, but not on $1/(1-x)^{1-4Nu}$. Thus in this case

$$w(x) \int_x \frac{d\xi}{V_{\delta \xi}w(\xi)}$$

is integrable near $x = 0$ if and only if

$$w(x) \int_x \frac{d\xi}{\xi}$$

is integrable, which is equal to

$$(1 - x)^{-4Nu} \int_x \frac{d\xi}{\xi} = -(1 - x)^{-4Nu} \log x .$$

This is certainly integrable in a neighborhood of $x = 0$, because

$$\int_x \log x = -x(1 - \log x) .$$

Hence

(3.18) $$w(x) \int_x \frac{d\xi}{V_{\delta\xi}w(\xi)} \in L(0, \tfrac{1}{2}) .$$

By (3.15) and (3.18), $r_0 = 0$ is exit.

Another example of a different type is provided by the process of a random environment, (Wright, 1948; Kimura, 1954). Kimura (1954) has given a Kolmogorov forward equation whose KBE is

$$\frac{\partial u}{\partial t} = \frac{\sigma^2 x^2 (1 - x)^2}{2} \frac{\partial^2 u}{\partial x^2} .$$

(Compare this KBE with (2.23).) Thus

$$w(x) = 1,$$

$$\frac{1}{V_{\delta x}w(x)} = \frac{(\sigma^2 + \bar{s})}{x^2(1 - x)^2} \notin L(0, \tfrac{1}{2}) ,$$

$$\notin L(\tfrac{1}{2}, 1) .$$

Therefore the boundaries are not regular. And

$$w(x) \int \frac{d\xi}{V_{\delta\xi}w(\xi)} \sim \frac{1}{x} \text{ or } \frac{1}{1 - x} \notin L(0, \tfrac{1}{2}) ,$$

$$\notin L(\tfrac{1}{2}, 1) .$$

Hence they are not exit boundaries. Finally,

$$\frac{1}{V_{\delta x}w(x)} \int^x w(\xi) = \frac{(\sigma^2 + \bar{s})}{x^2(1 - x)^2} \text{ or } \frac{(\sigma^2 + \bar{s})(1 - x)}{x^2(1 - x)^2}$$

$$= \frac{(\sigma^2 + \bar{s})}{x(1 - x)} \notin L(0, \tfrac{1}{2}) \text{ or } = \frac{(\sigma^2 + \bar{s})}{x^2(1 - x)} \notin L(\tfrac{1}{2}, 1) .$$

They can not be entrance boundaries. Therefore both $r_0 = 0$ and $r_1 = 1$ must be natural boundaries. For this process a trajectory can never reach 0 or 1.

CHAPTER 4

EXPECTATION OF INTEGRATION ALONG SAMPLE PATHS

4.1 Integration along sample paths

We have repeatedly emphasized the correspondence between a diffusion process and the appropriate KBE

(4.1) $$\frac{\partial u(t, x)}{\partial t} = Au(t, x) = \left\{ \frac{V_{\delta x}}{2} \frac{\partial^2}{\partial x^2} + M_{\delta x} \frac{\partial}{\partial x} \right\} u(t, x)$$

where

(4.2) $$u(t, x) = \int p(t, x, y) f(y) dy$$

with

(4.3) $$u(0, x) = f(x) .$$

The function f(y) in (4.2) can actually be any function. Thus u(t, x)
is the expectation of the quantity given by f(x) as a function of the
gene frequency x. Now consider the integral of u(t, x) from t = 0 to
t = ∞;

(4.4) $$F(x) \equiv \int_0^\infty u(t, x) dt .$$

Applying the operator A of (4.1) to F(x), we have

(4.5) $$AF(x) = \int_0^\infty Au(t, x) dt = \int_0^\infty \frac{\partial u(t, x)}{\partial t} dt$$

$$= u(\infty, x) - u(0, x) = u(\infty, x) - f(x) .$$

Therefore if the process has at least one exit, then $u(\infty, x) = 0$ for
all x,

44

(4.6)   AF(x) + f(x) = 0 .

These equations can be used in obtaining the total sum of u(t, x) over all t, i.e., $\int_0^\infty u(t, x)dt$. Symbolically, equation (4.6) can be solved by applying the inverse operator A^{-1} to both sides, i.e., $F(x) = -A^{-1}f(x)$. This method was first used in genetics by Kimura and Maruyama (1969), and then generalized by Maruyama and Kimura (1971).

   If the quantity of our interest is the probability that a path stays in an interval (x_0, x_1), then

(4.7)   f(x) = 1   for $x_0 < x < x_1$
                         = 0   otherwise

(see Fig. 4.1).
A special case of the above situation would be that x_0 and x_1 are the two boundaries of state space of the process. Another special case of (4.7) is the time in which a path stays in a small neighborhood of a given point, say y. Then

(4.8)   f(x) = δ(x - y)

where δ( ) is Dirac's delta function, (see Fig. 4.2).

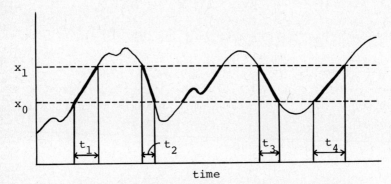

Fig. 4.1 Diagram illustrating a sample
path and the sojourn in (x_0, x_1),
$t_1+t_2+t_3+t_4$.

The thick lines in Fig. 4.1 indicate that the path is in (x_0, x_1). Since u(t, x) is the probability that a path is inside of (x_0, x_1) at time t, F(x) is the sum of the total duration of time when a path is inside of (x_0, x_1). With f(x) = δ(x - y), u(t, x) = probability density that a path is at y.

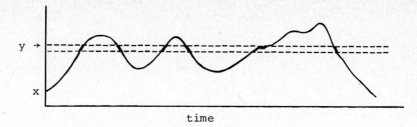

Fig. 4.2 Diagram illustrating a sample
path and the sojourn in $(y,\ y+dy)$.

Since this special case later plays a very important role in our discussions, we shall give it a special notation;

$$(4.9) \quad \Phi(x,\ y) = \int_0^\infty u(t,\ x)\,dt = \int_0^\infty \int p(t,\ x,\ \xi)\,\delta(y - \xi)\,d\xi\,dt.$$

The function $\Phi(x,\ y)$ of (4.9) is called Green's function or Green's
operator of the differential equation

$$AF(x) + f(x) = 0,$$

which has the following property: For an arbitrary $f(x)$,

$$(4.10) \qquad\qquad g(x) \equiv -\int \Phi(x,\ y)\,f(y)\,dy \equiv Gf(x)$$

is the solution of

$$Ag(x) + f(x) = 0\ .$$

Therefore the operator G defined in (4.10) acts as an inverse operator
of A. Therefore $G = A^{-1}$. Of course it also has a probablistic interpretation. Note that

$$\Phi(x,\ y) = \int_0^\infty u(t,\ x)\,dt$$

$$= \int_0^\infty \int_0^1 p(t,\ x,\ \xi)\,\delta(\xi - y)\,d\xi\,dt\ .$$

This is, therefore, the total "sojourn" time in a neighborhood of y,
given that a path starts from x. Hence

$$\int_0^1 \Phi(x,\ y)\,f(y)\,dy = \int_0^1 f(y)\int_0^\infty \int_0^1 p(t,\ x,\ \xi)\,\delta(y - \xi)\,d\xi\,dt\,dy$$

$$= \int_0^\infty \int_0^1 p(t, x, \xi) f(\xi) d\xi dt$$

$$= \int_0^\infty u(t, x) dt.$$

The mean value of the random variable X_t will be given by letting

(4.11) $f(x) = x$,

and the second moment of X_t by letting

(4.12) $f(x) = x^2$.

In the case of the genetic models, they correspond to the mean gene frequency and the second moment of it at time t, given that the frequency is x at t = 0. The number of homozygote individuals can be calculated by letting

(4.13) $f(x) = Nx^2$

and that of heterozygotes by letting

(4.14) $f(x) = 2Nx(1 - x)$.

The proportion of heterozygotes is of course given by letting

(4.15) $f(x) = 2x(1 - x)$.

(For interesting applications of (4.13),(4.14) and (4.15), see Nei, 1971; Li and Nei, 1972; Li, 1975.)

Starting from the position x, the probability that a path will reach r_0 within time t without previously passing through r_1, can be calculated by letting

$$f(r_0) = 1$$

$$f(x) = 0 \qquad \text{otherwise} .$$

We should again note that r_0 need not be the end point of the space.

4.2   The boundary conditions

The KBE (4.1) and the differential equations (4.5) or (4.6) may not have unique solutions, unless we specify the conditions at the boundary r_0 and r_1. As discussed before, if both boundaries are in-accessible, no arbitrary boundary condition can be imposed and there is only one process obeying KBE (4.1). On the other hand, if at least one of the boundaries is accessible, we can have various processes, obeying the KBE (4.1), and the distinction can be made only by the boundary conditions.

Absorbing and reflecting boundaries are two special cases of the general return process. In the case of an absorbing boundary, a path terminates at the first arrival of the path at the boundary. At a reflecting boundary, a path will be immediately returned into the interior of (r_0, r_1). For an absorbing boundary,

(4.16)
$$u(t, r_0) = 0 \qquad \text{for all } t,$$
and
$$F(r_0) = 0$$

which correspond the solutions of KBE (4.1) and equation (4.9) respectively. For a reflecting boundary the conditions are

(4.17)
$$\frac{\partial u(t, x)}{\partial x}\bigg|_{x=r_0} < 0,$$

$$\frac{dF(x)}{dx}\bigg|_{x=r_0} < 0.$$

4.3 An example

Consider the pure random drift case of constant population size N for which the KBE is

$$\frac{\partial u}{\partial t} = \frac{x(1-x)}{4N} \frac{\partial^2 u}{\partial x^2}.$$

First we shall compute the time, $T(x)$, required for a gene of initial frequency x to reach fixation or extinction by genetic random drift.

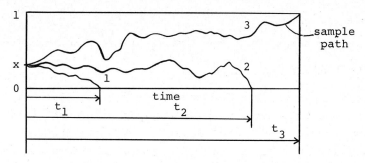

Fig. 4.3 Diagram illustrating sample paths going extinction (path 1 and 2) and fixation (path 3).

In Fig. 4.3, t_1 is the time for path 1, t_2 is for path 2, and so on. Function T(x) is the average time for all possible paths as a function of initial frequency x. Note that all paths start from x at t = 0, and also that paths going to fixation as well as those destined for extinction are included. In this case r_0 = 0 and r_1 = 1, and according to (4.7),

$$f(x) = 1 \qquad 0 < x < 1,$$

$$f(0) = f(1) = 0 .$$

Function, T(x), must satisfy equation (4.5) which in this case will be

$$(4.18) \qquad AT(x) + 1 = \frac{x(1 - x)}{4N} \frac{d^2T(x)}{dx^2} + 1 = 0$$

with T(0) = T(1) = 0, (see also Darling and Siegert, 1953; Feller, 1954). From (4.18),

$$\frac{d^2T(x)}{dx^2} = -4N \frac{1}{x(1 - x)} = -4N \left[\frac{1}{x} + \frac{1}{1 - x} \right] .$$

Hence

$$\frac{dT(x)}{dx^2} = -4N[x(1 - \log x) - \log(1 - x)] + a$$

and

$$T(x) = 4N[x(1 - \log x) + (1 - x)\{1 - \log(1 - x)\}] .$$
$$+ ax + b$$

Now

$$0 = T(0) = 4N + b \Rightarrow b = -4N$$

$$0 = T(1) = 4N + a - 4N \Rightarrow a = 0 .$$

Finally then, as Watterson (1962) showed,

$$(4.19) \qquad T(x) = -4N[x \log x + (1 - x)\log(1 - x)] .$$

Formula T(x) of (4.19) is the average time (in generations) required for an allele to become extinct from the population or to spread to the whole, given that its initial frequency is x. For example if x = 0.5,

$$(4.20) \qquad T(0.5) = -4N[\tfrac{1}{2} \log \tfrac{1}{2} + \tfrac{1}{2} \log \tfrac{1}{2}]$$

$$= 4N \log 2 = 4N \times 0.693 = 2.77N.$$

This means that if a path starts from x = 1/2, it will take on the

average about 2.8N generations to reach fixation or extinction. If
x = 1/2N which can be regarded as a case of a newly arisen mutant,

$$T(\tfrac{1}{2N}) = -4N \left[\tfrac{1}{2N} \log \tfrac{1}{2N} + \tfrac{2N - 1}{2N} \log(1 - \tfrac{1}{2N}) \right]$$

$$= -2 \left[\log \tfrac{1}{2N} + (2N - 1)\log(1 - \tfrac{1}{2N}) \right]$$

$$\doteq 2[\log 2N + \tfrac{2N - 1}{2N}]$$

$$\doteq 2[\log 2N + 1] .$$

If 2N is large,

(4.21) $$T(\tfrac{1}{2N}) \approx 2 \log 2N .$$

Note that the time given in (4.20) and (4.21) are dependent on the
population size N.

4.4 Green's function for a pure random process

Let us now obtain Green's function of this case. The equation to
be solved is, according to (4.5) and (4.8),

$$A\Phi(x, y) = \frac{x(1 - x)}{4N} \frac{d^2\Phi(x, y)}{dx^2} = -\delta(x - y) .$$

Hence

$$\frac{\partial\Phi(x, y)}{\partial x} = -4N \int_0^x \frac{\delta(\xi - y)}{\xi(1 - \xi)} d\xi + a$$

and

$$\Phi(x, y) = -4N \int_0^x \int_0^\theta \frac{\delta(\xi - y)}{\xi(1 - \xi)} d\xi d\theta + ax + b .$$

Now

$$\Phi(0, y) = 0 \implies b = 0$$

$$\Phi(1, y) = 0$$

$$= -4N \int_0^1 \int_0^\theta \frac{\delta(\xi - y)}{\xi(1 - \xi)} d\xi d\theta + a$$

$$= -4N \int_y^1 \frac{d\theta}{y(1 - y)} d\theta + a$$

$$= \frac{-4N}{y} + a = 0$$

thus

$$a = \frac{4N}{y}.$$

Therefore the final formula is

(4.22) $\qquad \Phi(x, y) = -4N \int_0^x \int_0^\theta \frac{\delta(\xi - y)}{\xi(1 - \xi)} \, d\xi d\theta + \frac{4Nx}{y}.$

Note that

$$\int_0^x \int_0^\theta \frac{\delta(\xi - y)}{\xi(1 - \xi)} \, d\xi d\theta = 0 \qquad \text{if } x < y,$$

$$= \frac{(x - y)}{y(1 - y)} \qquad \text{if } x > y.$$

Thus (4.22) can be simplified to

(4.23) $\quad \begin{cases} \Phi(x, y) = \dfrac{4Nx}{y} & \text{for } x < y \\[2mm] \Phi(x, y) = \dfrac{4Nx}{y} - \dfrac{4N(x - y)}{y(1 - y)} = \dfrac{4N(1 - x)}{1 - y} & \text{for } x > y. \end{cases}$

Let us first check this result by calculating T(x) of (4.19). According-
ing to the inverse operator argument given by (4.10), we must have

$$T(x) = \int_0^1 \Phi(x, y) \, dy$$

because f(x) = 1 for all x. But

(4.24) $\quad \displaystyle\int_0^1 \Phi(x, y) = \int_0^x \frac{4N(1 - x)}{1 - y} \, dy + \int_x^1 \frac{4Nx}{y} \, dy$

$$= 4N(1 - x)\log(1 - y) \Big|_x^0 + 4Nx \log x \Big|_x^1$$

$$= -4N[(1 - x)\log(1 - x) + x \log x]$$

which agrees with T(x) of (4.19). Note that the first term on the
right side of (4.24) is the average time spent in the area (0, x) and
the second term in the same formula is the time spent in the other
part (x, 1). Therefore, the two terms in T(x) of (4.19) correspond to
the two different parts in which a path stays, (see Fig. 4.4).

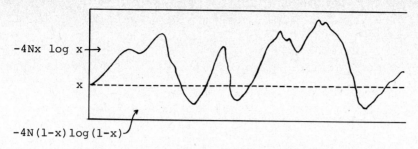

-4Nx log x →

x

-4N(1-x)log(1-x)

Fig. 4.4 Diagram illustrating a sample path
in the regions above x and below x.

Let us next calculate the time spent in an interval other than
the entire space (0, 1), say (0.1, 0.9). The appropriate solution denoted by T(x; 0.1, 0.9) should be given by

$$\int_0^1 \Phi(x, y) f(y) dy$$

where now f(y) = 1 for 0.1 < y < 0.9 and f(y) = 0 otherwise. Therefore

$$T(x; 0.1, 0.9) = \int_{0.1}^{0.9} \Phi(x, y) dy .$$

If 0.1 < x < 0.9,

$$T(x; 0.1, 0.9) = \int_{0.1}^{x} \frac{4N(1 - x)}{1 - y} dy + \int_{x}^{0.9} \frac{4Nx}{y} dy$$

$$= 4N(1 - x)[\log 0.9 - \log(1 - x)]$$
$$+ 4Nx[\log 0.9 - \log x]$$

$$= 4N \log 0.9 - 4Nx \log x$$
$$- 4N(1 - x)\log(1 - x);$$

if x < 0.1

$$(4.25) \qquad T(x; 0.1, 0.9) = \int_{0.1}^{0.9} \frac{4Nx}{y} dy = 4Nx \log 9;$$

and if x > 0.9, then

$$T(x; 0.1, 0.9) = \int_{0.1}^{0.9} \frac{4N(1 - x)}{1 - y} dy$$

$$= 4N(1 - x)\log 9 .$$

Assuming that N is large, we can compare $T(\frac{1}{2N})$ of (4.19) and

$T(\frac{1}{2N} ; 0.1, 0.9)$ of (4.25):

$$T(\frac{1}{2N}) \approx 2 \log 2N \quad \text{and}$$

$$T(\frac{1}{2N} ; 0.1, 0.9) = 2 \log 9 .$$

It is interesting to note that if $x_0 > 0$ and $x_1 < 1$, and if the population is large, the time spent in an interval (x_0, x_1) is independent of the population size N. For example, in the case of (0.1, 0.9), it is equal to 2 log 9.

For references on Green's function (operator) and the role of the inverse operator, see any textbook on either ordinary differential or partial differential equations, e.g., Coddington and Levinson (1955); Goursat (1915); Ince (1927); Courant and Hilbert (1962); Morse and Feshback (1953).

4.5 Computer simulation

The properties of stochastic processes can be demonstrated by computer simulations. The following example is to simulate the time to fixation or extinction, and the time a path sojourns in (0.5, 0.75). In the simulation, N = 50, x = 0.02, and pure random drift are assumed. The average results were printed out at every 100 repeats until 1000. This was done using UTHERCC system of the University of Texas at Houston. This is an exact copy of the program and the output:

```
        PROGRAM B(INPUT, OUTPUT, TAPE5 = INPUT, TAPE6 = OUTPUT)
C       SIMULATION ON TIME TO FIXATION OR EXTINCTION, AND TIME SPENT
C       IN AN INTERVAL, WRITTEN BY T. MARUYAMA AND RUN AT 'UTHERCC'.
C       THE FOLLOWING PARAMETERS ARE TO BE SET, BUT NO DATA ARE READ IN
C       N=POPULATION NUMBER, INT=INITIAL NUMBER OF MUTANTS, LBD=LOWER
C       BOUND OF THE INTERVAL, LAD=UPPER BOUND, INTVL=INTERVAL AT WHICH
C       PRINTOUT IS MADE, MAXM=MAXIMUM NUMBER OF REPETITIONS(RUNS).
C       RANF(DUD) IN THE PROGRAM IS A FUNCTION THAT GENERATES RANDOM
C       NUMBERS UNIFORMLY DISTRIBUTED IN (0,1), AND THAT IS USED FOR
C       SAMPLING OF GAMETES.
        MAXM=1000
        INTVL=100
        N=50
        INT=2
        NN=N+N
        LBD=N
        LAD=NN-(N/2)
        NNM=NN-1
        FNN=NN
        DUD=0.123456789
        NS=0
        AVRT=0.0
        BVRT=0.0
        IOUT=INTVL
```

```
            X0=FLOAT(INT)/FNN
            X1=FLOAT(LBD)/FNN
            X2=FLOAT(LAD)/FNN
            WRITE(6,70)N,X0,X1,X2
       70   FORMAT(1H1,20X,'RESULTS OF SIMULATION',//,20X,'POPULATION
                                                     SIZE =',
           1I3,10X,'INITIAL FREQUENCY =',F5.2,10X,//,20X,'NO. OF REPEATS',
                                                     '/',
           2'TIME TO FIX. EXT.','/','TIME SPENT BETWEEN',F4.2,1X,'AND',
                                                     F4.2)
       10   NS=NS+1
C           SETTING UP FIRST GENERATION
            NGN=INT
C           TESTING FIXATION OR EXTINCTION
      300   IF(NGN.LT.1) GO TO 100
            IF(NGN.EQ.NN) GO TO 100
            AVRT=AVRT+1.0
C           TESTING PATH IN THE INTERVAL
            IF(NGN.LT.LBD) GO TO 200
            IF(NGN.GT.LAD) GO TO 200
            BVRT=BVRT+1.0
C           SAMPLING OF GAMETES
      200   IBM=0
            FRQ=FLOAT(NGN)/FNN
            DO 20 I=1,NN
            X=RANF(DUD)
            IF(X-FRQ)30,20,20
       30   IBM=IBM+1
       20   CONTINUE
            NGN=IBM
            GO TO 30
      100   IF(NS-IOUT)10,400,400
      400   IOUT=IOUT+INTVL
C            CALCULATION OF AVERAGE TIMES
            AV=AVRT/FLOAT(NS)
            BV=BVRT/FLOAT(NS)
            WRITE(6,50)NS,AV,BV
       50   FORMAT(1H0,20X,I5,11X,F10.4,11X,F10.4)
            IF(NS-MAXM)10,60,60
C            CALCULATION OF THEORETICAL VALUES BASED ON THE DIFFUSION MODEL
       60   THTT=-4.0*FLOAT(N)*(X0*ALOG(X0)+(1.0-X0)*ALOG(1.0-X0))
            THTIN=4.0*FLOAT(N)*X0*ALOG(X2/X1)
            WRITE(6,500)THTT, THTIN
      500   FORMAT(1H0,///,26X,'THEORETICAL EXPECTATION OF THE TIME TO
                                                     FIX.',/
           1,31X,'OR EXT. BASED ON THE DIFFUSION MODEL=',F10.4,//,26X,
                                                     'THEORE
           2TICAL EXPECTATION OF THE TIME SPENT',/,31X,'IN THE INTERVAL='
                                                     ,F10
           3.4)
            STOP
            END
```

RESULTS OF SIMULATION

POPULATION SIZE = 50   INITIAL FREQUENCY = .02

NO. OF REPEATS/TIME TO FIX. EXT./TIME SPENT BETWEEN .50 AND .75

NO. OF REPEATS	TIME TO FIX. EXT.	TIME SPENT BETWEEN .50 AND .75
100	18.3200	1.7500
200	25.0600	2.7540
300	20.5000	2.0500
400	19.0925	1.9575
500	17.8960	1.6140
600	18.7583	1.7583
700	20.6957	1.9957
800	19.8350	1.8312
900	19.5178	1.6767
1000	19.3900	1.5990

THEORETICAL EXPECTATION OF THE TIME TO FIX.
    OR EXT. BASED ON THE DIFFUSION MODEL = 19.6078

THEORETICAL EXPECTATION OF THE TIME SPENT
    IN THE INTERVAL = 1.6219

4.6 Sum of heterozygotes

Using Green's function of the above problem, we can calculate the sum of the heterozygotes that appear in the whole history of a path. But we shall first calculate it using equation (4.5). According to (4.14).

$$AH(x) = -2Nx(1 - x)$$

where $H(x)$ is the sum of heterozygote individuals, given that the initial frequency is x. The above equation is

$$\frac{x(1 - x)}{4N} \frac{d^2 H(x)}{dx^2} = -2Nx(1 - x).$$

Hence, the appropriate solution with boundary conditions $H(0) = H(1) = 0$ is

(4.26) $$H(x) = 4N^2 x(1 - x).$$

In particualr, if $x = 1/(2N)$,

(4.27) $$H(\frac{1}{2N}) = 2N(1 - \frac{1}{2N}).$$

Now let us calculate the same quantity with Green's function. Recall that our Green's function is

$$\Phi(x, y) = \frac{4Nx}{y} \qquad x < y,$$

$$= \frac{4N(1 - x)}{1 - y} \qquad x > y.$$

Therefore

$$H(x) = \int_0^1 2Ny(1 - y)\, \Phi(x, y)\, dy$$

$$= 8N^2 x \int_x^1 \frac{y(1 - y)}{y} + 8N^2(1 - x) \int_0^x \frac{y(1 - y)}{1 - y}\, dy$$

$$= 8N^2(1 - x) \int_0^x y\, dy + 8N^2 x \int_x^1 (1 - y)\, dy$$

(4.28)
$$= 4N^2 x^2(1 - x) + 4N^2 x(1 - x)^2$$

(4.29)
$$= 4N^2 x(1 - x).$$

Formula (4.29) agrees with (4.26). However formula (4.28) provides more information: The first term is the sum of heterozygotes appearing while a path is in (0, x), and the second term is the sum while a path is in (x, 1).

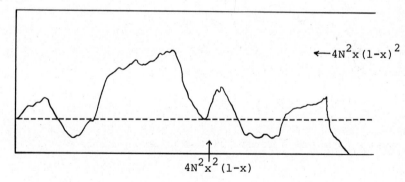

Fig. 4.5 Diagram illustrating a sample path and amounts of heterozygotes in the regions above x and below x.

4.7 Process with reflecting boundary

In the previous section, we have studied a case of absorbing boundaries at both ends. We shall next consider cases of return processes. The first example will be a case where a reflecting boundary

is constructed artificially. The second example is the less artificial reflection by mutation pressure.

   We shall first consider the pure random drift process whose KBE is

(4.30)
$$\frac{\partial u}{\partial t} = \frac{x(1-x)}{4N} \frac{\partial^2 u}{\partial x^2} .$$

Assume that if a path reaches x = 1, we place it immediately in the
interior of (0, 1). Let the place to be returned be a single location,
say x_1 = 1-ε. We assume that the other boundary x = 0 is absorbing.
Therefore if a path reaches fixation it is returned, but if it goes to
extinction, it vanishes. The process still satisfies the same KBE,
but now the boundary conditions are

(4.31)
$$u(t, 1) = u(t, x_1)$$

(4.32)
$$u(t, 0) = 0$$

where x_1 = 1-ε.

   Instead of solving this problem directly, let us consider a
slightly modified version. Suppose that the boundary of the process
at x = 1 is moved to x_1 = 1-ε. Note that the new boundary x_1 is not
an exit, but a regular boundary. Now let the process be reflecting at
this boundary. The KBE for this process is the same as (4.30), but
the boundary condition is

(4.33)
$$\left. \frac{\partial u(t, x)}{\partial x} \right|_{x=x_1} < \infty .$$

   Green's function is most informative and to obtain it, we need to
solve the equation

(4.34)
$$\frac{x(1-x)}{4N} \frac{\partial^2 \Phi(x, y)}{\partial x^2} = -\delta(x - y)$$

for 0 < x, y < x_1 and it must satisfy the boundary conditions

(4.35)
$$\Phi(0, y) = 0 ,$$

(4.36)
$$\left. \frac{\partial \Phi(x, y)}{\partial x} \right|_{x=x_1} = 0 .$$

Integrating (4.34) once, we have

(4.37)
$$\frac{\partial \Phi(x, y)}{\partial x} = -4N \int_0^x \frac{\delta(\xi - y)}{\xi(1 - \xi)} d\xi + a .$$

From (4.36)

$$\left. \frac{\partial \Phi(x, y)}{\partial x} \right|_{x=x_1} = \frac{-4N}{y(1 - y)} + a = 0$$

and so

$$a = \frac{4N}{y(1 - y)} \quad .$$

Integrating (4.37) in which the above value is substituted for a, we have

$$\Phi(x, y) = -4N \int_0^x \int_0^\theta \frac{\delta(\xi - y)}{\xi(1 - \xi)} + \frac{4Nx}{y(1 - y)} + b .$$

From (4.35), b = 0. Since

$$\int_0^x \int_0^\theta \frac{\delta(\xi - y)}{\xi(1 - \xi)} \, d\xi d\theta = 0 \qquad \text{if } x < y,$$

$$= \frac{(x - y)}{y(1 - y)} \qquad \text{if } x > y.$$

Green's function is

(4.38)
$$\Phi(x, y) = \frac{4Nx}{y(1 - y)} \qquad \text{for } x < y,$$

$$= \frac{4N}{1 - y} \qquad \text{for } x > y.$$

With the aid of this function, we can calculate the time a path stays in the interval $(0, x_1)$ before it reaches the absorbing boundary at $x = 0$. Let us denote it by $T(x, x_1)$. Then

$$T(x, x_1) = \int_0^{x_1} \Phi(x, y) dy = \int_0^x \frac{4N}{1 - y} \, dy + \int_x^{x_1} \frac{4Nx}{y(1 - y)} \, dy$$

(4.39)
$$= -4N \log(1 - x) + 4Nx \log \frac{x_1(1 - x)}{x(1 - x_1)} \quad .$$

The second term on the right side of (4.39) can be rewritten as

(4.40) $4Nx[\log x_1 + \log(1 - x_1) - \log x - \log(1 - x_1)]$.

The last term in (4.40), $\log(1 - x_1)$ diverges as x_1 approaches 1. Therefore, the location of the boundary near 1 affects the behavior of a path very much. Note that this process with a reflecting boundary

is almost the same as the process proposed at the begining of this
section, in which a path is returned from x = 1 to x_1 = 1-ε instantaneously. Hence the location of the return point makes a large difference in this case of the sojourn time.

However there are quantities which are almost independent of the
return point x_1 (or the reflecting point x_1), and which are associated
with the same process. One such quantity of interest is the heterozygosity. Let $H(x, x_1)$ be the average amount of heterozygosity produced in the whole history of a path. Then

$$H(x, x_1) = \int_0^{x_1} 2y(1 - y)\Phi(x, y)\,dy$$

$$= 4N \int_0^x \frac{2y(1 - y)}{1 - y}\,dy + 4Nx \int_x^{x_1} \frac{2y(1 - y)}{y(1 - y)}\,dy .$$

$$H(x, x_1) = 4Nx(2x_1 - x) .$$

Note that since x_1 = 1-ε

$$H(x, 1-\varepsilon) = 4Nx(2 - x - 2\varepsilon)$$

and

(4.41) $$H(x, 1) = 4Nx(2 - x) .$$

Therefore the heterozygosity associated with this return process is
not significantly affected when the boundary approaches 1. The limiting case is given in (4.41). In particular if the initial frequency
is x = 1/2N

$$H(\frac{1}{2N}, 1) = 2(2 - \frac{1}{2N}) \approx 4 ,$$

whereas the corresponding value for the case with two absorbing boundaries at x = 0 and x = 1 is 2. It is interesting that the heterozygosities in the two situations differ by a factor of 2.

From the nature of Green's function given in (4.38) it is clear
that the quantity which vanishes as y approaches 1 will not be affected
by the location of the return point. In such a case we can impose the
boundary condition

$$\left. \frac{\partial u(t, x)}{\partial x} \right|_{x=1} < \infty .$$

Let us check this with the formula obtained by solving directly the corresponding differential equation,

$$\frac{x(1-x)}{4N} \frac{\partial^2 H(x)}{\partial x^2} = -2x(1-x) ,$$

with the boundary conditions

$$\frac{\partial H(1)}{\partial x} = 0 \quad \text{and} \quad H(0) = 0 .$$

Hence the appropriate solution is

$$H(x) = 4Nx(2-x) .$$

This agrees with formula (4.41).

4.8 Irreversible mutation model (or infinite alleles)

This is a more natural case involving one reflecting boundary. The KBE is

(4.42) $\qquad \dfrac{\partial u(t, x)}{\partial t} = \dfrac{x(1-x)}{4N} \dfrac{\partial^2 u(t, x)}{\partial x^2} - ux \dfrac{\partial u(t, x)}{\partial x} .$

It has been shown that the boundary at $x = 0$ is an exit, and the boundary at $x = 1$ is an entrance if $4Nu > 1$ and it is regular if $4Nu < 1$.

We shall first examine the entrance case. The boundary condition would be

(4.43) $\qquad \left. \dfrac{u(t, x)}{\partial x} \right|_{x=1} < \infty.$

Green's function of this case satisfies

(4.44) $\qquad \dfrac{x(1-x)}{4N} \dfrac{\partial^2 \Phi(x, y)}{dx^2} - ux \dfrac{d\Phi(x, y)}{dx} + \delta(x - y) = 0 ,$

and the appropriate solution is given by

(4.46) $\quad \Phi(x, y) = \dfrac{4N\left\{1 - (1-x)^{1-4Nu}\right\}}{(4Nu - 1)y(1-y)^{1-4Nu}} \qquad \text{for } x < y,$

$\qquad\qquad = \dfrac{4N\left\{1 - (1-x)^{1-4Nu}\right\}}{(4Nu - 1)y(1-y)^{1-4Nu}} + \dfrac{4N\left\{(1-y)^{4Nu} - (1-x)^{4Nu}\right\}}{(1 - 4Nu)y(1-y)^{1-4Nu}}$

$\qquad\qquad\qquad\qquad\qquad\qquad\qquad\qquad\qquad\qquad \text{for } x > y.$

As x gets small

$$(1-x)^{1-4Nu} \approx 1 - (1 - 4Nu)x .$$

Thus for small x, formula (4.46) reduces to

(4.47) $$\Phi(x, y) \approx 4Nxy^{-1}(1 - y)^{1-4Nu}.$$

In particular if $x = 1/2N$

(4.48) $$\Phi(\tfrac{1}{2N}, y) = 2y^{-1}(1 - y)^{4Nu-1}, \qquad y \geq \tfrac{1}{2N}.$$

This formula gives the average time (in generations) spent in a neighborhood of y, before a path reaches extinction, given that it starts from $x = 1/2N$. More precisely $\Phi(\tfrac{1}{2N}, y)dy$ is the average time spent in the interval $(y-dy/2, y+dy/2)$.

If the population is in equilibrium, then 2Nu new mutant genes will be introduced into the population each generation, while the same number of mutants will become extinct in each generation. Then the ergodic theory, which asserts that the time average is equal to the space average, applies to the situation and the (expected) gene frequency distribution is

(4.49) $$2Nu\Phi(\tfrac{1}{2N}, y) = 4Nuy^{-1}(1 - y)^{4Nu-1}.$$

The reason for taking $x = 1/2N$ is that each mutant is unique. Formula (4.49) was first derived by Wright (1948), and then by Kimura and Crow (1964), and by Ewens (1969). Returning to Green's function $\Phi(x, y)$ given in (4.46), note that the derivation of the function does not depend on the assumption that $x = 1$ is an entrance boundary (4Nu > 1). Therefore $\Phi(x, y)$ of (4.46) is valid also for $0 < 4Nu < 1$ in which $x = 1$ is a regular boundary.

Assuming that 4Nu < 1 and therefore that $x = 1$ is accessible, we can impose an absorbing boundary at $x = 1$. Then Green's function for this case is again the solution of (4.44) with the boundary conditions

(4.50) $$\Phi(0, y) = \Phi(1, y) = 0.$$

The solution is

(4.51) $$\Phi(x, y) = \frac{4N}{1 - 4Nu}\left\{1 - (1 - x)^{1-4Nu}\right\}/ y \qquad \text{for } x < y,$$

$$= \frac{4N(1 - x)}{1 - 4Nu}\left\{\frac{(1 - x)^{1-4Nu}}{y(1 - y)^{1-4Nu}} - (1 - x)^{1-4Nu}\right\} \text{ for } x > y.$$

The function $\Phi(x, y)dy$ is the average time that a path starting from x stays in the interval $(y-dy/2, y+dy/2)$ before it reaches either $x = 1$ or $x = 0$.

A case of particular interest is

$$(4.52) \qquad \Phi(\frac{1}{2N}, y) = \frac{4N}{1 - 4Nu}\left\{1 - (1 - \frac{1}{2N})^{1-4Nu}\right\}/y \approx \frac{2}{y}$$

$$\text{for } y > x = \frac{1}{2N}.$$

It is interesting to note that the last expression in (4.52) is independent of 4Nu, though it is required that 4Nu < 1. Therefore, before the first arrival at x = 1 or x = 0, the behavior of a path for this process is independent of the mutation pressure. It looks like a paradox. Recall Green's function for the pure random drift process. The upper formula of (4.23) is relevant to the present issue and it is

$$\Phi(x, y) = \frac{4Nx}{y}.$$

Therefore

$$\Phi(\frac{1}{2N}, y) = \frac{2}{y} \qquad \text{for } y > x = \frac{1}{2N}.$$

Indeed the above formula agrees with the last part of (4.52).

Using special cases (x = 1/2N) of two Green's functions for absorbing and reflecting boundaries at x = 1, the probability that a path of present frequency y has never reached fixation (x = 1) in the past can be obtained. Here we assume that every mutant is unique and the population is in equilibrium. Then the probability is equal to

$$(4.53) \qquad \frac{\Phi^{abs}(\frac{1}{2N}, y)}{\Phi^{refl}(\frac{1}{2N}, y)} = (1 - y)^{1-4Nu}$$

provided 1-4Nu > 0 which is the condition that x = 1 is accessible, (Maruyama, 1972). If 4Nu tends to 0, the probability becomes (1-y), and if 4Nu tends to 1, it becomes 1.

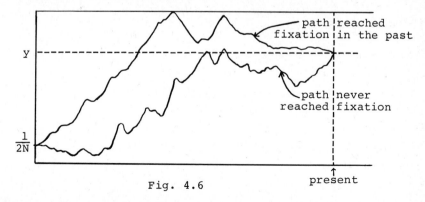

Fig. 4.6

The values of formula (4.53) can be easily obtained. Some numerical values for realistic 4Nu are presented in Table 4.1.

Table 4.1 Probability that an allele of present
frequency y has never been frequency 1 in the past.

4Nu \ y	0.01	0.1	0.2	0.5	0.8	0.9	0.99
0.1	0.991	0.910	0.818	0.536	0.235	0.126	0.016
0.2	0.992	0.919	0.837	0.574	0.276	0.158	0.025
0.5	0.995	0.949	0.894	0.707	0.447	0.316	0.100
0.9	0.999	0.990	0.978	0.933	0.851	0.794	0.631

4.9 General form of Green's function

So far we have dealt with the sum of a particular quantity associated with the whole history of a path, or with a part of the history in which the path stayed in a specified gene frequency range. Such a sum can be obtained by

$$\int_E \Phi(x, y) f(y) dy$$

in which E indicates integration over the area under consideration.

We shall now obtain general formulae. Consider the general form of the KBE

(4.54)
$$\frac{\partial u}{\partial t} = \frac{V_{\delta x}}{2} \frac{\partial^2 u}{\partial x^2} + M_{\delta x} \frac{\partial u}{\partial x}$$

and the boundaries are $r_0 < r_1$. Denote by $\Phi_{abs}^{abs}(x, y)$, $\Phi_{abs}^{refl}(x, y)$ and $\Phi_{refl}^{abs}(x, y)$, Green's functions of the processes with absorbing boundary at r_0 and absorbing at r_1, with absorbing at r_0 and reflecting at r_1 and with reflecting at r_0 and absorbing at r_1, respectively. All these functions satisfy the same KBE

(4.55)
$$\left\{ \frac{V_{\delta x}}{2} \frac{\partial^2}{\partial x^2} + M_{\delta x} \frac{\partial}{\partial x} \right\} \Phi(x, y) + \delta(x - y) = 0.$$

Upon solving this equation with appropriate boundary conditions, the final formulas are

(4.56)
$$\Phi_{abs}^{abs}(x, y) = \frac{2g(y, r_1)u(x)}{V_{\delta y}G(y)} \qquad x < y,$$

$$= \frac{2}{V_{\delta y} G(y)} \{g(y, r_1) u(x) - g(y, x)\} \quad x > y;$$

(4.57)
$$\Phi_{refl}^{abs}(x, y) = \frac{2g(y, r_1)}{V_{\delta y} G(y)} \quad x < y,$$

$$= \frac{2}{V_{\delta y} G(y)} \{g(y, r_1) - g(y, x)\} \quad x > y;$$

and

(4.58)
$$\Phi_{abs}^{refl}(x, y) = \frac{2g(r_0, x)}{V_{\delta y} G(y)} \quad x < y$$

$$= \frac{2}{V_{\delta y} G(y)} \{g(r_0, x) - g(y, x)\} \quad x > y$$

where

(4.59)
$$G(x) = \exp \left\{ -2 \int_{\gamma_0}^{x} \frac{M_{\delta\xi}}{V_{\delta\xi}} d\xi \right\},$$

(4.60)
$$g(a, b) = \int_{a}^{b} G(x) dx$$

and

(4.61)
$$u(x) = g(r_0, x)/g(r_0, r_1) .$$

The formulae (4.56) $\sim$ (4.58) are valid only if they exist.

4.10 Probability of fixation

A subject that has not been discussed in detail yet in these notes, but is important in population genetics, is the fixation probability and its generalization. Suppose that if a path arrives at the boundary r_0 without previously reaching the other boundary we shall give it a value, say ξ_0, and once a path reaches either boundary it is ignored. Similarly if it arrives at r_1 without reaching r_0, we shall give it a value, say ξ_1. We are interested in the expectation of the values, given that every path starts from x. As discussed before, the expectation denoted by $u(x)$ satisfies

$$\left\{ \frac{V_{\delta x}}{2} \frac{d^2}{dx^2} + M_{\delta x} \frac{d}{dx} \right\} u(x) = -f(x)$$

where $f(x) = 0$ for $r_0 < x < r_1$ and the boundary conditions are

$$u(r_0) = \xi_0$$

and

$$u(r_1) = \xi_1 .$$

The appropriate solution is of course

(4.62)
$$u(x) = \frac{(\xi_1 - \xi_0)g(r_0, x)}{g(r_0, r_1)} + \xi_0$$

where $g(a, b)$ is given in (4.60). If r_0 is inaccessible, it may happen that

$$u(x) = \xi_1 , \qquad \text{for all } x ,$$

and if r_1 is inaccessible, it is possible that

$$u(x) = \xi_0 \qquad \text{for all } x .$$

In the case of $\xi_0 = 0$ and $\xi_1 = 1$, $u(x)$ is the probability that a path starting from x reaches r_1 without previously arriving at r_0. If $\xi_0 = 1$ and $\xi_1 = 0$, $u(x)$ is the arrival probability at r_0. Note that r_0 and r_1 need not be real boundaries of the state space. In the case of genetic problems, we usually use

$$0 \le r_0 \le r_1 \le 1 .$$

The ultimate fixation probability of a mutant is given by the solution of (4.62) with $\xi_0 = 0$ and $\xi_1 = 1$, and therefore

(4.63)
$$u(x) = \frac{g(0, x)}{g(0, 1)} .$$

(Kimura, 1962). If we use the selection model (2.22) where one gene has relative fitness 1+s and where the effects of genes are additive,

(4.64)
$$u(x) = \frac{1 - e^{-4Nsx}}{1 - e^{-4Ns}} .$$

If $0 < 4Nsx << 1$ and $4Ns >> 1$

$$u(x) \approx 4Nsx$$

and if $x = 1/2N$,

(4.65)
$$u(x) \approx 2s .$$

Therefore a singly present mutant with advantage s has a chance of 2s to spread in the whole population. For instance, a mutant of 1% advantage will be fixed with a probability of 2%. (See Haldane, 1927; Kimura, 1957, also Feller 1954.)

4.11 Behavior of sample paths near the origin

It is of our interest to inquire into the chance of a mutant reaching a certain frequency, say $y_1 \ll 1$, without going to extinction. The above theory can be used for this problem. If the fitnesses are additive and the initial frequency is 1/2N, which means that a single copy of the mutant is present, then the probability that the frequency reaches y_1 before extinction is given by

$$(4.66) \qquad \frac{1 - e^{-2s}}{1 - e^{-4Nsy_1}} \approx \frac{2s}{1 - e^{-4Nsy_1}} .$$

Hence if $|4Nsy_1|$ is small, this probability is approximately equal to

$$(4.67) \qquad \frac{2s}{4Nsy_1} = \frac{1}{2Ny_1} .$$

This is also equal to the probability that a single, neutral mutant reaches frequency y_1. Thus even if $|4Ns|$ is large, a mutant behaves very much like a neutral allele near y_1 if y_1 is small. It is interesting to compare (4.66) with the ultimate fixation probability which is given by

$$(4.68) \qquad \frac{1 - e^{-2s}}{1 - e^{-4Ns}} \approx \frac{2s}{1 - e^{-4Ns}} \qquad \text{for } s \ll 1 .$$

This formula asserts that for a mutant to behave like a neutral allele for the whole range of frequency, $|4Ns|$ must be small. Therefore the criteria for the two quantities can be different by a factor of 10 ($y_1 = 0.1$), or even 100 ($y_1 = 0.01$). Some examples are shown in Table 4.2.

Essentially the same is true for the sojourn time in a neighborhood of y, whose density in the case of genic selection is equal to

$$(4.69) \qquad \frac{2}{y} \quad (s = 0) \qquad \text{or}$$

$$(4.70) \qquad \frac{(1 - e^{-4Ns(1-y)})(1 - e^{-2s})}{y(1 - y)s(1 - e^{-4Ns})} \qquad (s \neq 0) .$$

Table 4.2 Probability to reach a certain frequency
and of fixation. (Initially one gene)

y_1 4Ns	Probability to reach y_1			Ultimate fixation probability
	0.01	0.05	0.1	
-100	0.00030	0.0000034	10^{-8}	1.85×10^{-48}
-10	0.00047	0.000077	0.000029	2.27×10^{-9}
-5	0.00049	0.000088	0.000039	1.69×10^{-7}
-1	0.00050	0.000098	0.000048	2.91×10^{-6}
0	0.00050	0.000103	0.000053	7.91×10^{-6}
5	0.00051	0.000113	0.000064	2.52×10^{-5}
10	0.00053	0.000127	0.000079	5×10^{-5}
100	0.00079	0.000503	0.000500	5×10^{-4}

For a range where $|4Nsy|$ is small, the last formula reduces to $2/y$
which is the same as (4.69) for $s = 0$. It is important to note that
the criterion is the value of $|4Nsy|$, but not of $|4Ns|$. Therefore
irrespective of its effect, a mutant behaves like a neutral allele in
a range of the gene frequency y for which $|4Nsy|$ is small. If we take
the ratio of (4.69) to (4.70), we have

$$1 - y \qquad \text{for small y and } 4Ns \gg 0,$$

and

$$1 - 4Nsy \qquad \text{for small 4Nsy and } 4Ns \ll 0.$$

Some cases of sojourn times are given in Table 4.3. It is appearent
from the table that, if $|4Nsy|$ is small, the behavior of a sample path
in the neighborhood of y is nearly independent of the selection pressure. Using this property, Nei (1977) has derived a powerful formula
that enables us to estimate the mutation rate in natural populations
from the distribution of low frequency genes.

Table 4.3 Sojourn time of a sample path in range (y, y+1/2N), given x_0 = 1/2N, n = 10000 and genic selection (s).

4Ns	y = 0.001	y = 0.005	y = 0.01	y = 0.05	y = 0.1
-100	0.0906	0.0122	3.72×10^{-3}	1.42×10^{-5}	5.00×10^{-8}
-10	0.0991	0.0191	9.15×10^{-3}	1.28×10^{-3}	4.08×10^{-4}
-5	0.0996	0.0196	9.6×10^{-3}	1.64×10^{-3}	6.74×10^{-4}
-1	0.0998	0.0199	9.94×10^{-3}	1.94×10^{-3}	9.44×10^{-4}
0	0.1000	0.0200	0.0100	2×10^{-3}	10^{-3}
1	0.1000	0.0200	0.0100	2.04×10^{-3}	1.04×10^{-3}
10	0.1001	0.0201	0.0101	2.105×10^{-3}	1.111×10^{-3}
100	0.1001	0.0201	0.1010	2.105×10^{-3}	1.111×10^{-3}

4.12 Higher moments

So far we have discussed the expectation (the first moment in statistical terms) of a quantity which is associated with a process. We will next seek a method for the second or higher moments of the quantity. We start with notation and definitions. Denote a sample path (trajectory) by ω, and the time at which path ω is disregarded from our attention by τ_ω. For example, if we are interested in a path until it reaches fixation, then τ_ω is the first time that the sample path ω reaches fixation. We should be aware of that τ_ω depends on the sample path ω, and therefore different paths may have different values of τ_ω. We denote by $x_{t,\omega}$ the value of the random variable in path ω at time t.

Now for an arbitrary function f(x), consider the integral

(4.71) $$\int_0^{\tau_\omega} f(x_\xi, \omega)\, d\xi .$$

This is the sum of the quantity given by f(x) along the path ω, from the position at time 0 until the path ω reaches the point where it is disregarded from our attention. Since τ_ω is path dependent, the integral (4.71) also depends on path ω. Now we take the expectation of the integral over the collection of all paths which start from x, i.e., x_0 = x, and let

(4.72) $$F(x) = E_x \left\{ \int_0^{\tau_\omega} f(x_\xi, \omega)\, d\xi \right\}$$

where E{ } indicates the expectation and x indicates the initial condition. In fact, the function F(x) satisfies

$$F(x) = \int \Phi(x, y) f(y) dy$$

and it is the solution of

(4.73) $AF(x) + f(x) = 0$.

Using the integral (4.71), we can consider

$$e^{-\lambda \int_0^{\tau_\omega} f(x_\xi, \omega) d\xi}$$

where λ is a parameter, and take the expectation of this quantity over the collection of all paths starting from the same initial point at time 0. Let

(4.74) $\phi(\lambda, x) = E_x \left\{ e^{-\lambda \int_0^{\tau_\omega} f(x_{\xi\omega}) d\xi} \right\}$

$$= E_x \left\{ 1 - \lambda \int_0^{\tau_\omega} f(x_\xi, \omega) d\xi + \frac{1}{2} \left(-\lambda \int_0^{\tau_\omega} f(x_{\xi\omega}) d\xi \right)^2 \right.$$

$$\left. + \frac{1}{3!} \left(-\lambda \int_0^{\tau_\omega} f(x_\xi, \omega) d\xi \right)^3 + \cdots \right\}.$$

It can be shown that function $\phi(\lambda, x)$ of (4.74) satisfies

(4.75) $A\phi(\lambda, x) - \lambda f(x) \phi(\lambda, x) = 0$

where

$$A = \frac{V_{\delta x}}{2} \frac{d^2}{dx^2} + M_{\delta x} \frac{d}{dx}.$$

If we differentiate (4.75) once with respect to λ, and then let $\lambda = 0$, we have

$$AF(x) + f(x) = 0,$$

which is the equation given in (4.73). This is the equation for the expectation. If we differentiate (4.75) n times with respect to λ and then let $\lambda = 0$, we have

(4.76) $AF_n(x) + nf(x) F_{n-1}(x) = 0$

where

$$F_n(x) = E_x \left\{ \left(\int_0^{\tau_\omega} f(x_\xi, \omega) \, d\xi \right)^n \right\} ,$$

$$F_0(x) = 1 .$$

This function gives the n-th moment of the quantity in question among all possible sample paths from a given initial point. Equation (4.76) is an iteration, and using this we can obtain successively higher moments.

A special case of (4.72) would be

(4.77) $$F(x) = E_x \left\{ f(x_{\tau_\omega}, \omega) \right\} .$$

We are interested in the value of $f(x)$ when path ω leaves the area of our interest, which here is the value of $f(x)$ at the boundary. Then

(4.78) $$AF(x) = 0$$

and $F(x)$ assumes preassigned values at the boundaries. If $f(0) = 0$ and $f(1) = 1$, this is equivalent to giving a value of zero if a path reaches frequency 0 and value 1 if a path reaches frequency 1. Therefore the solution of (4.78) with $f(0) = 0$ and $f(1) = 1$ is the ultimate fixation probability of a mutant with initial frequency x. This is the problem discussed in section 4.10.

From a mathematical view point, (4.76) is an inhomogeneous equation, while (4.78) is a homogeneous equation.

Summarizing, the function (or the n-th moment)

(4.79) $$F_n(x) = E_x \left\{ \left(\int_0^{\tau_\omega} f(x_\xi, \omega) \, d\xi \right)^n \right\}$$

satisfies

(4.80) $$AF_n(x) + nf(x) F_{n-1}(x) = 0$$

$$\text{with} \quad F_0(x) = 1 ,$$

and

$$F(x) = E_x \left\{ f(x_{\tau_\omega}, \omega) \right\}$$

satisfies

$$AF(x) = 0$$

with preassigned value at the boundaries.

4.13 Nagylaki's formula

In order to obtain a higher moment of a quantity defined by (4.79), we need to solve equation (4.80) recursively. That method, therefore, can become quite cumbersome for a higher moment. Nagylaki (1974) has overcome this difficulty and shown that the higher moments defined by (4.79) can be obtained more conveniently as functions of the mean sojourn time functions (4.56) $\sim$ (4.58).

To do this we first rewrite $F_n(x)$ of (4.79) as

$$(4.81) \qquad F_n(x) = E\left\{\prod_{i=1}^{n} \int_0^{\tau_\omega} f(x_{\xi_i}, \omega)\, d\xi_i\right\}.$$

Nagylaki (1974) argues as follows. Since n! permutations of the time ξ_i contribute equally to $F_n(x)$, we can order them so that $\xi_i \geq \xi_{i-1}$. Then equation (4.81) becomes

$$F_n(x) = n! \prod_{i=1}^{n} \int_0^{\xi_{i+1}} d\xi_i \int_0^1 dx_i\, f(x_i)\, \phi(\xi_{i-1} - \xi_i, x_{i-1}, x_i)$$

where $\xi_0 = 0$, $\xi_{n+1} = \infty$, $x_0 = x$ and $\phi(t, x, y)$ is the transition density of the process. Now if we let $\xi_i' = \xi_i - \xi_{i-1}$ $(\xi_i = \sum_{j=1}^{i} \xi_j')$,

$$(4.82) \qquad F_n(x) = n! \prod_{i=1}^{n} \int_0^1 dx_i \int_0^\infty d\xi_i'\, f(x_i)\, \phi(\xi_i', x_{i-1}, x_i)\ .$$

Noting

$$\int_0^\infty \phi(\xi, x, y)\, d\xi = \Phi(x, y)\ ,$$

$$(4.83) \qquad F_n(x) = n! \prod_{i=1}^{n} \int_0^1 dx_i\, f(x_i)\, \Phi(x_{i-1}, x_i)\ .$$

Therefore, the n-th moment can be expressed as a function of the mean sojourn time.

Nagylaki (1974) used formula to obtain the sojourn time distribution. Substituting $f(x) = \delta(x-y)$ into (4.83), we have the n-th moment of the sojourn time,

$$F_n(x) = n! \prod_{i=1}^{n} \int_0^1 dx_i\, \delta(x_i - y)\, \Phi(x_{i-1}, x_i)$$

$$= n!\, \Phi(x, y)\, [\Phi(y, y)]^{n-1}.$$

Therefore the moment generating function ought to be

$$(4.84) \qquad 1 + \frac{\lambda \Phi(x, y)}{1 - \lambda \Phi(y, y)} = \int_0^\infty e^{\lambda t} p(x, y; t) \, dt$$

where $p(x, y; t) \, dy$ is the probability that a sample path starting from x sojourns t generations in the gene frequency range $(y, y+dy)$. Inverting the moment generating function (4.84), which is the Laplace transform of $p(x, y; t)$, we have

$$(4.85) \qquad p(x, y; t) = [1 - \frac{\Phi(x, y)}{\Phi(y, y)}]\delta(t) + \frac{\Phi(x, y)}{\Phi(y, y)^2} \exp(- \frac{t}{\Phi(y, y)}) .$$

Problem 1. Consider an equilibrium population of size N and a number of loci of similar nature. Assume that mutation rate per gene per generation is u and furthermore that 2Nu is sufficiently small so that practically every mutant occurs at a "homallelic" locus. By homallelic or "monomorphic" we mean that there is only one type of allele at the locus in question. We assume that all mutants are selectively neutral. (See Kimura (1969).)

(i) What fraction of the loci on the average will be "polymorphic", if we call a locus polymorphic when it has more than one type of allele segregating?

(ii) What fraction will be polymorphic, if we define polymorphic to be a locus that has at least one allele whose frequency is less than 99% but greater than 1%? (Assume 2N > 100).

(iii) What is the average heterozygosity? The heterozygosity is defined here as the probability that two randomly chosen homologous genes are different types.

(iv) What is the distribution of loci which give a fixed amount of heterozygosity?

(v) How long (how many generations) does a newly arising mutant stay in the frequency range (0.5, 1), before it reaches either fixation or extinction?

(vi) How long does it stay in the same range (0.5, 1), if it starts from frequency x = 0.5?

Problem 2. For the infinite allele model, show that the average number of different alleles segregating at a locus, in an equilibrium population of size N, is equal to

$$4Nu \int_{\frac{1}{2N}}^{1} x^{-1}(1 - x)^{4Nu-1} dx.$$

Problem 3.   Consider the selection model whose KBE is

$$\frac{\partial u}{\partial t} = \frac{x(1-x)}{4N} \frac{\partial^2 u}{\partial x^2} + sx(1-x)\frac{\partial u}{\partial x} .$$

i) Obtain Green's function by solving

$$\left\{ \frac{x(1-x)}{4N} \frac{\partial^2}{\partial x^2} + sx(1-x)\frac{\partial}{\partial x} \right\} \Phi(x, y) + \delta(x-y) = 0 ,$$

$$\text{with } \Phi(0, y) = \Phi(1, y) = 0 .$$

(ii)   Calculate the average of the total amount of heterozygosity produced in the whole history of a path.

(iii)   Obtain the probability that a mutant of initial frequency x reaches fixation before it becomes extinct, (the ultimate fixation probability).

(iv)   Obtain the probability that a mutant of initial frequency x reaches the gene frequency 0.5, before it becomes extinct.

(v)   Assume that s < 0 and that there is no possibility of the mutant reaching fixation. Try to calculate the number of individuals who will be a carrier of the deleterious genes. Calculate the homozygous and heterozygous carriers separately (Li and Nei, 1972).

Problem 4.   Suppose that there are two unlinked loci and that at one locus, a single selectively neutral allele is introduced, and at the other locus, a single selectively advantageous mutant is introduced. Compare the total heterozygosities, due to the two mutant genes, which will appear before the mutants become either fixed or extinct. Assume that Ns >> 0 but s is small.

Problem 5.   Consider the model of the changing population size given by KBE (2.13). Calculate the time required for a mutant of initial frequency x to become either fixed or extinct, on two assumptions:

$$\text{(i)} \quad N_t = N_0(1+rt) ,$$

$$\text{(ii)} \quad N_t = N_0 e^{rt}$$

where r is a constant. Does a path eventually reach either the frequency 0 or the frequency 1, with probability 1 in case (ii)?

Problem 6.   Consider Kimura's model of a random environment given by the KBE,

$$\frac{\partial u(t, x)}{\partial t} = \frac{\sigma^2 x^2 (1 - x)^2}{2} \frac{\partial^2 u(t, x)}{\partial x^2} .$$

Calculate the time required for a path starting from $x = 0.5$ to reach either $x = 1-\delta$ or $x = \delta$, where δ is a fixed, small number. As δ tends to zero, does the first arrival time approach an exponential function of δ? (Reference Kimura, 1954).

Problem 7.   Consider the model of a random environment and assume that the population size is finite (N). Then the variance $V_{\delta x}$ has an additional term due to sampling of gametes, which is equal to

$$\frac{x(1 - x)}{2N} .$$

Therefore the total variance is

$$V_{\delta x} = \sigma^2 x^2 (1 - x)^2 + \frac{x(1 - x)}{2N}$$

and the KBE is

$$\frac{\partial u}{\partial t} = \frac{1}{2} \left\{ \sigma^2 x^2 (1 - x)^2 + \frac{x(1 - x)}{2N} \right\} \frac{\partial^2 u}{\partial x^2} .$$

(i)   Show that $x = 1$ and $x = 0$ are now exit boundaries, instead of natural.

(ii)   Obtain Green's function.

(iii)   Obtain the average time to either fixation or extinction.

(iv)   Obtain the average total heterozygosity.

Problem 8.   Consider a large number of populations of equal size. Assume that each population contains two alleles, say A_1 and A_2, and that the frequency of the A_1 gene is uniformly distributed in (0, 1), among the populations.

(i)   Calculate the average time for a randomly chosen population to reach either fixation or extinction of the A_1 gene.

(ii)   Note that the average gene frequency among the populations is 0.5. Compare the average time obtained in (i) with $T(\frac{1}{2})$ of formula (4.20).

CHAPTER 5

MODIFICATION OF PROCESSES

In the preceding chapter, we have dealt with a given process and
obtained formulae for the expectations of various quantities in question, summed along the whole history of a path. In those discussions,
we made no distinction among paths of different destinations, nor
allowed extinction or creation of new paths before it reached a certain
boundary. We shall now introduce some modifications, such as selection
of only those paths of a fixed destination, and show that the same
methods used in the previous discussions apply to the new processes.
The only change needed is to modify the infinitesimal operator by an
appropriate function. The modifications to be discussed are:
(i) Killing and creating of paths in the interior of the state space.
(ii) Selection of paths with a fixed destination.
(iii) Change of time parameter along a path.
    We should note that the three modifications are not to change the
"road map" of the paths of the original processes. Throughout this
chapter, we assume that there is an original process given by the KBE

(5.1) $$\frac{\partial u}{\partial t} = Au = \left\{ \frac{V_{\delta x}}{2} \frac{\partial^2}{\partial x^2} + M_{\delta x} \frac{\partial}{\partial x} \right\} u .$$

Reference: Dynkin (1965).

5.1 Killing and creating paths

    Suppose that a path has a certain probability that it will vanish
or create a new copy, and that the killing and birth of a path is a
function of the location. (Time can enter too.) We assume that a new
path starts from scratch at the location of its birth. Let $b(x)\Delta t$ be
the probability that a path at x will create a new copy in time Δt,

and let $d(x) \Delta t$ be the probability of extinction for a path at x. Then let $c(x) \equiv b(x) - d(x)$. The new process with the birth and death possibility of a path has the following infinitesimal operator

(5.2) $$\tilde{A} = A + c(x)$$

where A is the operator of the original process. Therefore the KBE for the new process is

$$\frac{\partial u}{\partial t} = \left\{ \frac{V_{\delta x}}{2} \frac{\partial^2}{\partial x^2} + M_{\delta x} \frac{\partial}{\partial x} + c(x) \right\} u.$$

For example, consider a hypothetical situation in which the A_1 gene is selectively advantageous within a population, but it causes some possibility of extinction of the population itself. Let s be the selective advantage of the A_1 gene within a population and let rx be the probability of extinction of the population if its A_1-gene frequency is x. Then the operator of the new process is

$$\tilde{A} = \frac{x(1 - x)}{4N} \frac{\partial^2}{\partial x^2} + sx(1 - x)\frac{\partial}{\partial x} - rx.$$

5.2 Selection of paths

Let $q(x)$ be a solution of either

(5.3) $$Aq(x) = 0$$

or

(5.3') $$Aq(x) + f(x) = 0$$

where A is the operator of the original process and $f(x)$ is an arbitrary function. Let $p(t, x, y)$ be the transition probability density of the original process. Then let

(5.4) $$\tilde{p}(t, x, y) \equiv \frac{1}{q(x)} \int_\Omega p(t, x, y) q(y) \, dy.$$

It can be shown that the operator of the new process governed by the $\tilde{p}(t, x, y)$ is

(5.5) $$\tilde{A} = \frac{1}{q(x)} Aq(x).$$

As in (4.76), the n-th moment of a quantity in question, along the paths of the new process, satisfies

(5.6)
$$\tilde{A}F_n(x) + nf(x)F_{n-1}(x)$$

$$= \left\{ \frac{1}{q(x)} \, Aq(x) \right\} F_n(x) + nf(x)F_{n-1}(x) = 0$$

which can be rewritten as

(5.7)
$$Aq(x)F_n(x) + nf(x)q(x)F_{n-1}(x) = 0 .$$

The more convenient form of (5.7) would be

(5.8)
$$AG_n(x) + nf(x)G_{n-1}(x) = 0$$

where $G_n(x) = q(x)F_n(x)$.

Nagylaki's formula (4.83) holds for a higher moment of the new process, if we replace Green's function $\Phi(x_{i-1}, x_i)$ by that of the new process.

If we choose $q(x)$ to be the solution of $Aq(x) = 0$ with $q(r_0) = 0$ and $q(r_1) = 1$, then $q(x) = u(x)$ is the fixation probability of a path starting from x. The most important case is the fixation probability of a mutant gene in the whole population, which is given by the solution of (5.3) with boundary conditions $q(0) = 0$ and $q(1) = 1$. (See Fig. 5.1).

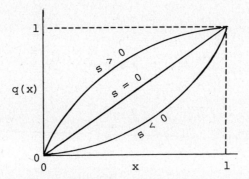

Fig. 5.1 Fixation probability as the
function of initial gene frequency.
Advantageous if s > 0 and disadvan-
tageous if s < 0.

With this choice of $q(x)$, the modification is to select only those paths destined to fixation. The n-th moment of a quantity of our interest along those paths leading to fixation can be obtained by solving

(5.9)
$$\tilde{A}F_n(x) + nf(x)F_{n-1}(x) = 0$$

which is

(5.10)
$$\frac{1}{u(x)} Au(x)F_n(x) + nf(x)F_{n-1}(x) = 0 .$$

As in (5.8), letting $G_n(x) \equiv u(x)F_n(x)$, we have, from (5.10),

(5.11)
$$AG_n(x) + nf(x)G_n(x) = 0$$

with $G_0(x) = u(x)$.

5.3 Random drift and fixation time

Consider the random drift case and let us calculate the average time required for a mutant gene to reach fixation, (excluding those paths leading to extinction). Recall that for the model and this problem

$$A = \frac{x(1 - x)}{4N} \frac{\partial^2}{\partial x^2} ,$$

$$u(x) = x$$

and

$$f(x) = 1 ,$$

Let $t_1(x) = F_1(x) = G_1(x)/u(x)$ be the average time for fixation. Substituting these quantities in (5.7), we have

(5.12)
$$\frac{x(1 - x)}{4Nx} \frac{d^2G_1(x)}{dx^2} + x = 0 .$$

Hence (omitting the subscript)

$$G''(x) = -4N \frac{1}{1 - x}$$

and so

$$G(x) = 4N\{(1 - x) - (1 - x)\log(1 - x)\} + ax + b .$$

Note

$$G(0) = 4N + b = 0$$

and

$$G(1) = a + b = 0$$

so that

$$a = -b = 4N .$$

Hence we have

$$G(x) = 4N(1 - x) - 4N(1 - x)\log(1 - x) + 4N(x - 1)$$

$$= -4N(1 - x)\log(1 - x)$$

and

(5.13) $$t_1(x) = \frac{G(x)}{x} = -4N\frac{1 - x}{x}\log(1 - x) .$$

As x tends 0,

(5.14) $$\lim_{x \to 0} t_1(x) = 4N.$$

This is a well know result of Kimura and Ohta (1969). As x tends to 1, say $x = (2N-1)/2N = 1 - 1/2N$, as shown by Kimura and Ohta (1969a),

(5.15) $$t_1(1 - \frac{1}{2N}) = -4N(\frac{2N}{2N - 1} - 1)\log\frac{1}{2N}$$

$$\approx 2 \log 2N.$$

Noting a symmetry in going from $x = 1 - \frac{1}{2N}$ to $x = 1$ and going from $x = \frac{1}{2N}$ to $x = 0$, formula (5.15) gives the average time for a mutant of frequency $1/2N$ to become extinct, given that it never reaches fixation.

If we let $\Phi_1(x, y)$ be Green's function of the new process (i.e., looking only at paths destined to fixation), then it satisfies

(5.16) $$\frac{1}{u(x)} Au(x)\Phi_1(x, y) + \delta(x - y) = 0.$$

If we let $G(x, y) \equiv u(x)\Phi(x, y)$, then (5.16) becomes

(5.17) $$AG(x, y) + u(x)\delta(x - y) = 0.$$

The special case of pure random drift $(u(x) = x)$ is

$$\frac{x(1 - x)}{4N} \frac{d^2G(x, y)}{dx^2} + x\delta(x - y) = 0.$$

Hence

$$G''(x, y) = \frac{-4N\delta(x - y)}{1 - x}$$

and so

$$G(x, y) = -4N \int_0^x \int_0^\theta \frac{\delta(\xi - y)}{1 - \xi} d\xi d\theta + ax + b.$$

Note $G(0, y) = 0$, $\Phi(0, y) = 0$ and $G(1, y) = 0$. Thus $b = 0$, $a = 4N$, and

$$G(x, y) = 4Nx \qquad \text{for } x < y,$$

$$= \frac{4Ny(1 - x)}{1 - y} \qquad \text{for } x > y.$$

Therefore

(5.18) $\qquad \Phi_1(x, y) = \dfrac{G(x, y)}{x} = 4N \qquad \text{for } x < y,$

$$= \frac{4Ny(1 - x)}{x(1 - y)} \qquad \text{for } x > y.$$

It is interesting to note from (5.18) that a path destined to fixation stays, on the average, for 2 generations at each of the intermediate gene frequency ranges, $(y, y+\frac{1}{2N})$ where $y = \frac{1}{2N}, \frac{2}{2N}, \frac{3}{2N}, \cdots, \frac{2N-1}{2N}$.

Formula (5.14) can be confirmed by using Green's function (5.18),

$$\int_0^1 \Phi_1(\frac{1}{2N}, y) dy = 4N.$$

(Here we ignored the range below $x = 1/2N$.)

5.4 A case of genic selection

Consider the case of genic selection governed by the KBE (2.22)

$$\frac{\partial u}{\partial t} = \frac{x(1 - x)}{4N} \frac{\partial^2 u}{\partial x^2} + sx(1 - x)\frac{\partial u}{\partial x}.$$

The probability of ultimate fixation is

$$u(x) = \frac{1 - e^{-4Nsx}}{1 - e^{-4Ns}}.$$

Using this $u(x)$, we can obtain Green's function for the genic selection case. That is

(5.19) $\quad \Phi_1(x, y) = \dfrac{(1 - e^{-4Nsy} - e^{-4Ns(1-y)} + e^{-4Ns})}{sy(1 - y)(1 - e^{-4Ns})} \qquad \text{for } x < y.$

The other part, for x > y, is more complicated. Note that multiplying both denominator and numerater of (5.19) by exp(4Ns) and rearranging,

$$\Phi_1(x, y) = \frac{e^{4Ns}}{e^{4Ns}} \Phi_1(x, y)$$

$$= \frac{(1 - e^{4Nsy} - e^{4Ns(1-y)} + e^{4Ns})}{-sy(1 - y)(1 - e^{4Ns})} .$$

This formula is Green's function for the case of -s. Therefore $\Phi_1(x,y)$ is independent of the sign of s, but depends on its absolute value $|s|$, provided y > x. We should also remember that we are looking at only those paths destined to fixation. Green's function for a range of x > y does depend on the sign of s, as well as its absolute value. Another property of $\Phi_1(x, y)$ of (5.19) is its symmetry about y = 1/2

$$\Phi_1(x, y) = \Phi_1(x, 1 - y), \qquad y > x .$$

In addition, $\Phi_1(x, y)$ assumes its minimum value at y = 1/2, provided x < 1/2.

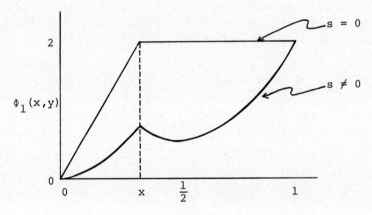

Fig. 5.2 Sojourn time of sample paths which start from frequency x and go to fixation.

5.5 A symmetric property of sample paths

There is a general property concerning the sojourn time in going from frequency x to y (> x) on the way to fixation and in going the reverse direction on the way to extinction, (Maruyama and Kimura, 1974). By solving (5.16) we have

(5.19) $\qquad \Phi_1(p, y) = 2u(y)\{1 - u(y)\}/\{v_{\delta y}u'(y)\}$

for $y \geq p$, and

(5.20) $\qquad \Phi_1(p, y) = 2\{1 - u(y)\}\{u(y)\}^2/\{u(p)v_{\delta y}u'(y)\}$

for $y \leq p$, where $u(y)$ is the fixation probability given by (4.62) and $u'(y)$ is its derivative. If we let $\Phi_0(q,y)$ be Green's function for the process starting from q and going to extinction, then similarly we have

(5.21) $\qquad \Phi_0(q, y) = 2u(q)\{1 - u(y)\}^2/[v_{\delta y}u'(y)\{1 - u(q)\}]$

for $y \geq q$, and

(5.22) $\qquad \Phi_0(q, y) = 2u(y)\{1 - u(y)\}/\{v_{\delta y}u'(y)\}$

for $y \leq q$. Let us suppose that $p < q$ and consider the behavior of the allele in the interval (p, q). Since $\Phi_1(p, y)$ of (5.19) and $\Phi_0(q, y)$ of (5.22) have the same expression and they are independent of p and q, the paths starting from p and going to fixation, and the paths starting from q and going to extinction have the identical mean sojourn time in the interval (p, q). In these two formulae, (5.19) and (5.22), if we let p go to 0 and q go to 1, then we have the situations where a single mutant gene spreads into the entire population and where the allele frequency decreases to zero from unity. Remarkably the features of the both processes are the same, despite the possible great difference in the probabilities of occurrence of these two events.

We can similarly obtain the higher moments of these quantities, using the above method. For example, the second moments denoted by $\Phi_1^{(2)}(p, y)$ and $\Phi_0^{(2)}(q, y)$, of the sojourn times are

$$\Phi_1^{(2)}(p, y) = \Phi_0^{(2)}(q, y) = 8[u(y)\{1 - u(y)\}]^2/\{v_{\delta y}u'(y)\}^2$$

for $p \leq y \leq q$.

5.6 A general formula

The general form of (5.17) is

(5.23) $\qquad AG_n(x) + u(x)F_{n-1}(x)f(x) = 0$

where

$$G_n(x) = u(x)F_n(x)$$

and

$$F_n(x) = E_x \left\{ \left[\int_0^{\tau_\omega} f(x_\xi, \omega) d\xi \right]^n \ \middle| \ x_{\tau_\omega} = 1 \right\} .$$

The solution is

$$G_n(x) = u(x)F_n(x)$$

$$= [1 - u(x)] \int_0^x \phi_{n-1}(\xi) u(\xi) d\xi + u(x) \int_0^1 \phi_{n-1}(\xi)[1 - u(\xi)] d\xi$$

where

$$\phi_n(x) = \frac{2Nf(x)F_{n-1}(x)g(0, 1)}{V_{\delta x}G(x)} ,$$

$$u(x) = g(0, x)/g(0, 1) ,$$

$$G(x) = \exp \left\{ -2 \int_0^{x} \frac{M_{\delta\xi}}{V_{\delta\xi}} d\xi \right\}$$

and

$$g(a, b) = \int_a^b G(x) dx .$$

References: Kimura and Ohta (1969); Maruyama and Kimura (1971);
Maruyama and Kimura (1975).

5.7 Age of sample paths

Another important case of the function $q(x)$ seems to be Green's
function of the original process, i.e., the solution of

(5.24) $$A\Phi(x, y) + \delta(x - y) = 0 .$$

With this choice of $q(x) = \Phi(x, y)$, we will examine paths which pass
through y at least once and until they leave y for the last time.
With the aid of this transformation, we can investigate the history of
a path in equilibrium given that it is in (y-dy/2, y+dy/2) at present.
For example, if we are interested in estimating the age of a path (the
number of generations the path has persisted in the population), given
that we know the present frequency and the initial frequency, then the
operator is

$$\tilde{A} = \frac{1}{\Phi(x, y)} A\Phi(x, y)$$

and

$$f(x) = 1.$$

If we let $B(x, y)$ be the age of an allele whose frequency is y, then it satisfies

(5.25)
$$\tilde{A}B(x, y) + 1 = 0$$

which is

(5.26)
$$AG(x, y) + \Phi(x, y) = 0$$

with

$$G(x, y) = \Phi(x, y)B(x, y) .$$

The particular case of the infinite neutral allele model, the average age of a mutant whose present frequency is y, is given by

(5.27)
$$Ag(y) = \frac{4N}{F}\left[\int_0^y \frac{\{1 - (1 - \xi)^F\}}{\xi(1 - \xi)^F} d\xi \right.$$
$$\left. + \{1 - (1 - y)^F\}\int_y^1 \frac{d\xi}{\xi(1 - \xi)^F}\right]$$

where

$$F = 1 - 4Nu .$$

(Remember that in this model we assumed that every mutant is unique and therefore every allele starts out from the frequency $1/2N$.) The value of the average age given by formula (5.27) includes the possibility that a path reached fixation in the past. In formula (5.27) for this situation, the $\Phi(x, y)$ is chosen to satisfy the boundary condition

$$\Phi(0, y) = \left.\frac{d\Phi(x, y)}{dx}\right|_{x=1} = 0 .$$

It is also of some interest to know the average age of a mutant assuming it has not reached fixation. For this problem, we have to solve (5.26) with the $\Phi(x, y)$ satisfying the boundary conditions $\Phi(0, y) = \Phi(1, y) = 0$. The formula is

(5.28)
$$Ag(y) = \frac{4N}{F}\left[\int_0^y \frac{\{1 - (1-\xi)^F\}}{\xi} d\xi + \frac{\{1 - (1-y)^F\}}{(1-y)^F}\int_y^1 \frac{(1-\xi)^F}{\xi} d\xi\right]$$

where $F = 1-4Nu$. The value given by formula (5.28) is the average age of a mutant that has frequency y and has not been fixed in the population.

The general method described above can be applied to a situation in which one of the alleles at the locus in question is selectively different from the rest. We will obtain an explicit formula for the age before fixation for an allele with an additive effect on fitness. Let s be the selection coefficient of the gene and ignore the subsequent mutation of this allele. Then $V_{\delta x} = x(1-x)/2N$ and $M_{\delta x} = sx(1-x)$. Solving first equation (5.24) for $\Phi(x, y)$ with the boundary conditions $\Phi(0, y) = \Phi(1, y) = 0$ and then equation (5.26) with this $\Phi(x, y)$, we have

$$(5.29) \quad Ag(y) = \frac{4N}{S(1 - e^{-S})} \int_0^1 \frac{(e^{S\xi} - 1)(e^{-S\xi} - e^{-S})}{\xi(1 - \xi)} d\xi$$

$$- \frac{4N}{S(e^{-Sy} - e^{-S})} \int_y^1 \frac{(1 - e^{-S(1-\xi)})(e^{-Sy} - e^{-S\xi})}{\xi(1 - \xi)} d\xi$$

where $S = 4Ns$. Note that this formula is independent of the sign of $S = 4Ns$. Thus the average age of an additive gene is independent of the direction of selection pressure.

Numerical values of formulae (5.27), (5.28) and (5.29) are tabulated for wide ranges of values of 4Nu and 4Ns, (Tables 5.1 ∿ 5.3).

The tables give the values of $Ag(y)/4N$. Therefore the actual average age can be calculated by multiplying the value by 4N. For instance if the present frequency of a mutant allele in question is 1% (0.01), 4Nu = 1 and the population size (N) is 10^5, then the average age calculated from Table 5.1 is $0.0557 \times 4 \times 10^5 = 22280$, about 20,000 generations.

The tables reveal several biologically interesting facts about the age including fixation. When 4Nu << 1, the age of an allele at high frequency is approximately equal to the reciprocal of the mutation rate, and the age of a low frequency allele is about y/u, where u is the mutation rate and y is the frequency. On the other hand, when 4Nu >> 1, the age tends to be much less. The values in Table 5.2 show that the age before fixation increases as the mutation rate increases. For example, the fixation time is 4N generations if u = 0, but it is $1.6 \times 4N$ generations if 4Nu = 1. The age before fixation therefore increases as 4Nu increases to 1 and then decreases as 4Nu becomes larger. References: Kimura and Ohta (1973) and Maruyama (1974).

Table 5.1 Numerical values of Ag(y)/4N of (5.27), where Ag(y) is the average age of a neutral gene whose frequency is y and which may have been fixed in the past. N is the population size.

y	4Nu									
	20	10	5	2	1	0.5	0.2	0.1	0.05	0.01
0.001	0.0040	0.0047	0.0054	0.0065	0.0074	0.0089	0.0123	0.0176	0.0278	0.1080
0.01	0.0228	0.0289	0.0357	0.0460	0.0557	0.0697	0.1042	0.1569	0.2588	1.0608
0.03	0.0448	0.0604	0.0787	0.1080	0.1361	0.1778	0.2807	0.4389	0.7446	3.1504
0.1	0.0827	0.1207	0.1697	0.2553	0.3432	0.4785	0.8189	1.3450	2.3635	10.3821
0.2	0.1105	0.1692	0.2500	0.4019	0.5666	0.8297	1.5047	2.5547	4.5903	20.6264
0.3	0.1284	0.2019	0.3074	0.5155	0.7507	1.1364	2.1422	3.7144	6.7661	30.8186
0.5	0.1523	0.2471	0.3904	0.6926	1.0553	1.6755	3.3341	5.9464	11.0277	51.1106
0.7	0.1686	0.2788	0.4512	0.8317	1.3089	2.1530	4.4560	8.1039	15.2115	71.3214
0.9	0.1810	0.3033	0.4995	0.9477	1.5298	2.5919	5.5334	10.2126	19.3428	91.4554
0.999	0.1864	0.3140	0.5204	0.9993	1.6302	2.8051	6.0614	11.2478	21.3762	101.4366

Table 5.2 Numerical values of Ag(y)/4N of (5.28),
where Ag(y) is the average age of a neutral gene
before fixation whose frequency is y, and where
N is the population size. (The values for 4Nu > 1
are identical with those in Table 5.1)

				4Nu			
y	1	0.9	0.7	0.5	0.3	0.1	0
0.001	0.0078	0.0077	0.0074	0.0072	0.0071	0.0069	0.0069
0.01	0.0561	0.0544	0.0521	0.0499	0.0482	0.0456	0.0450
0.03	0.1366	0.1321	0.1248	0.1177	0.1140	0.1086	0.1068
0.1	0.3418	0.3306	0.3074	0.2873	0.2704	0.2593	0.2536
0.2	0.5669	0.5426	0.4987	0.4623	0.4321	0.4105	0.4000
0.3	0.7528	0.7160	0.6529	0.6017	0.5595	0.5283	0.5136
0.5	1.0616	1.0010	0.9025	0.8245	0.7634	0.7129	0.6908
0.7	1.3203	1.2362	1.1059	1.0038	0.9240	0.8585	0.8298
0.9	1.5500	1.4403	1.2803	1.1561	1.0594	0.9804	0.9460
0.999	1.6458	1.5330	1.3591	1.2267	1.1219	1.0366	0.9995

Table 5.3 Numerical values of Ag(y)/4N of (5.29),
where Ag(y) is the average age of an additive
gene before fixation whose frequency is y.

				|4Ns|			
y	100	50	20	10	5	2	1	0
0.001	0.0024	0.0031	0.0040	0.0047	0.0055	0.0062	0.0064	0.0065
0.01	0.0113	0.0158	0.0230	0.0295	0.0368	0.0436	0.0453	0.0460
0.03	0.0191	0.0287	0.0458	0.0625	0.0824	0.1014	0.1061	0.1080
0.1	0.0302	0.0487	0.0866	0.1288	0.1837	0.2372	0.2504	0.2553
0.2	0.0379	0.0634	0.1198	0.1882	0.2820	0.3721	0.3937	0.4019
0.3	0.0432	0.0737	0.1442	0.2344	0.3607	0.4778	0.5053	0.5155
0.5	0.0516	0.0904	0.1857	0.3164	0.4946	0.6461	0.6801	0.6926
0.7	0.0601	0.1078	0.2312	0.4028	0.6147	0.7817	0.8183	0.8317
0.9	0.0744	0.1380	0.3024	0.5018	0.7258	0.8969	0.9341	0.9477
0.999	0.1027	0.1780	0.3513	0.5526	0.7772	0.9484	0.9857	0.9992

5.8 Number of affected individuals and genetic load

In the above discussion we are interested in the time spent between the start and the last exit from y, and therefore $f(x) = 1$ has
appeared repeatedly. Another case of our interest seems to be the
genetic load and the number of affected individuals in the past of a
path.

For the genetic load in the genic selection model,

(5.22) $f(x) = sx \qquad s < 0$.

For other genetic loads, see Crow and Kimura (1970). For the number
of affected individuals:

(5.23) $f(x) = N\{x^2 + 2x(1 - x)\}$;

which can be broken up into the number of homozygotes, for which

$$f(x) = Nx^2 ,$$

and the number of heterozygotes, for which

$$f(x) = 2Nx(1 - x) .$$

With these $f(x)$ functions, we solve the equation

(5.24) $AG(x, y) + f(x)\Phi(x, y) = 0$

where $G(x, y) = \Phi(x, y)F(x, y)$ and

$$F(x, y) = E_x \left\{ \int_0^{\tau_\omega} f(x_\xi, \omega)d\xi \,\middle|\, x_{\tau_y^\omega} = y \right\} .$$

The function $F(x, y)$ is the sum of the quantity, $f(x)$, associated with
the past of a path that is presently in a neighborhood of y. This is
a reversed question of what we have discussed in chapter 4, in the
sense that we ask the past of a path, instead of its future. Of course,
these quantities can be also interpreted in terms of the future of a
path. Namely, $F(x, y)$ will give the sum of the quantity associated
with a path on the condition that it starts from x and it is y at the
time of observation in the future.

5.9 Sojourn time of conditional sample paths on the present frequency

As in all previous cases, Green's function $\Phi(x, y, z)$ of this conditional process can be obtained. Green's function $\Phi(x, y, z)dz$, which is the time that a path spends in $(z-dz/2, z+dz/2)$, given that it starts from x and is presently at y , satisfies

(5.25) $$\frac{1}{\Phi(x, y)} A\Phi(x, y)\Phi(x, y, z) + \delta(x - z) = 0 ,$$

where $\Phi(x, y)$ is Green's function of the unconditional process. The solution of this equation for the pure random drift case is

(5.26) $$\Phi(x, y, z) = \frac{4Nx\Phi(z, y)}{z\Phi(x, y)} \qquad \text{for } x < z,$$

$$= \frac{4Nx\Phi(z, y)}{z\Phi(x, y)} - \frac{4N(x - z)\Phi(z, y)}{z(1 - z)\Phi(x, y)} \qquad \text{for } x > z.$$

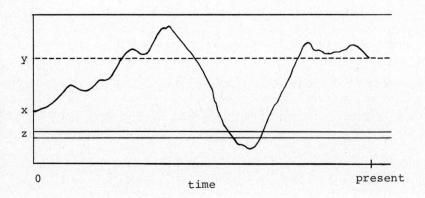

Fig. 5.3 Illustration of a sample path and the conditional sojourn time. x = initial frequency, y = present frequency.

With x = 1/2N,

(5.27) $$\Phi(\frac{1}{2N}, y, z) = 4N \qquad \text{for } z < y,$$

$$= \frac{4Ny(1 - z)}{(1 - y)z} \qquad \text{for } z > y.$$

It is interesting to note that in this case for z < y, where y is the present frequency, a path stays an average of 2 generations at each intermediate frequency prior to present, and for z > y, a path behaves as if it starts from z and is destined to fixation, (see the lower part of formula (5.18) and let z = x, then it become the same as the lower part of the above $\Phi(\frac{1}{2N}, y, z)$).

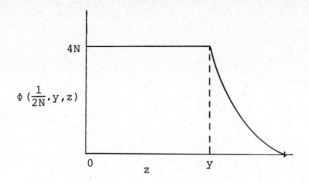

Fig. 5.4 Conditional sojourn time for
sample paths starting from 1/2N and
presently at y.

5.10 A conditional age

Suppose that we are interested in the paths that have not reached
either fixation or extinction. The conditional probability in this
case shall be similar to that of section 5.7, but much weaker in the
sense that we do not wish to restrict a path to be at any specific
location at the time of observation, requiring it only to be in the
interior of (0, 1). Then the q(x) to be used must be the solution of

(5.28) $$AT(x) + 1 = 0 .$$

Recall that this T(x) is the average time that a path stays inside of
the interior, i.e., (0, 1). Let

(5.29) $$F(x) = E_x \left\{ \int_0^t f(x_\xi, \omega) d\xi \,\Big|\, t < \tau_\omega \right\}$$

in which τ_ω is the "exit time" of path ω from (0, 1). Remember also
that in the case of a reflecting boundary at x = 1, a path can exit
only from x = 0. Then G(x) = T(x)F(x) satisfies

(5.30) $$AG(x) + f(x)T(x) = 0 .$$

In the case of the average age of selectively neutral genes segregating in a population of size N,

$$A = \frac{x(1-x)}{4N} \frac{d^2}{dx^2} ,$$

$$f(x) = 1 ,$$

and

$$q(x) = T(x) = -4N[x \log x + (1 - x)\log(1 - x)]$$

(see formula (4.19)). Here we ignored the mutation pressure on the dynamics of a path. For this case $G(x) = T(x)F(x)$ is the solution of

$$\frac{x(1 - x)}{4N} \frac{d^2 G(x)}{dx^2} = 4N\{x \log x + (1 - x)\log(1 - x)\}$$

or

$$G''(x) = (4N)^2 \left\{ \frac{\log x}{(1 - x)} + \frac{\log(1 - x)}{x} \right\}.$$

The solution of this equation with boundary conditions $G(0) = G(1) = 0$ is

$$G(x) = (4N)^2 \left[\int_0^x \frac{\log \xi}{1 - \xi} \, d\xi + x \log x - (1 - x)\log(1 - x) + x\frac{\pi^2}{6} \right].$$

Hence the average age of segregating genes, under the assumption that every mutant starts from $1/2N$ and is introduced at a homallelic locus, is

(5.31)
$$\frac{G(\frac{1}{2N})}{T(\frac{1}{2N})} = \frac{2N\pi^2}{3 \log 2N}$$

this result can be checked by calculating the mean age more directly. Using Green's function for the case with absorbing boundaries at $x = 0$ and $x = 1$, we can obtain the average age of a mutant gene before reaching either fixation or extinction. The age for a mutant starting from $x = 1/2N$ is

(5.32)
$$\frac{-4Ny \log y}{1 - y}$$

where y is the present frequency. From a previous study we know also that a path starting from $x = 1/2N$ stays for

$$(\frac{4N}{y} \, x) \frac{1}{2N} = (\frac{2}{y}) \frac{1}{2N}$$

generations in $(y, y+1/2N)$, (see 4.23). And we know that the time to either fixation or extinction is

$$T(\frac{1}{2N}) \approx 2 \log 2N, \text{ (see (4.24))}.$$

Therefore the average age of a segregating mutant is

$$\frac{1}{T(\frac{1}{2N})} \int_0^1 \frac{-4Ny \log y}{1-y} \frac{y}{2} \, dy$$

$$= \frac{-4N}{\log 2N} \int_0^1 \frac{\log y}{1-y} \, dy = \frac{4N}{\log 2N} \frac{\pi^2}{6} = \frac{2N\pi^2}{3 \log 2N}$$

which agrees with formula (5.31).

Green's function of the conditional process, $\phi(x, y, T(x))$, gives the time that a path spends in $(y-dy/2, y+dy/2)$, given that it starts from x and it has not yet reached τ_ω which is the first exit time. It satisfies

$$\frac{1}{T(x)} AT(x) \phi(x, y, T(x)) + \delta(x - y) = 0.$$

In the case of pure random drift, the equation is

$$\frac{1}{T(x)} \frac{x(1-x)}{4N} \frac{d^2}{dx^2} [T(x)\phi] + \delta(x - y) = 0$$

and the solution with $T(0)\phi = T(1)\phi = 0$ is

$$T(x)\phi(x, y, T(x)) = \frac{4NxT(y)}{y} - \frac{4N(x-y)T(y)}{y(1-y)} \int_0^x \delta(\xi - y) \, d\xi$$

where $\int_0^x \delta(\xi - y) \, d\xi = 1$ if $x > y$ and $= 0$ if $x < y$.

In particular, if $x = 1/2N$,

$$\phi(\frac{1}{2N}, y, T(\frac{1}{2N})) = \frac{-4N}{-y \log 2N} [y \log y + (1 - y) \log(1 - y)]$$

$$= \frac{1}{-y \log 2N} T(y).$$

5.11   Computer simulation

The following is a computer simulation on the sum of heterozygosity along a path that has started from 2/2N and is presently at $y = 0.1, 0.15, 0.20, \cdots, 1.00$. The simulation results (HET SIM) were compared with theoretical expectation (HET THR). This was done by UTHERCC at University of Texas at Houston.

```
C          SIMULATION ON THE SUM OF HETEROZYGOSITY, 2X(1-X), ALONG A PATH
C          STARTING FROM X AND AT Y AT PRESENT
C          THE FOLLOWING PARAMETERS MUST BE SET, N=POPULATION SIZE, IN=
C          INITIAL NUMBER OF GENES, MAX=MAXIMUM NUMBER OF REPEATS,
C          INTVL=INTERVAL AT WHICH OUT PUT IS PRINTED
           DIMENSION H(100),TT(100),V(100),TH(100)
           N=10
           IN=2
           MAX=2000
           NN=N+N
           INTVL=100
           NS=0
           IOUT=INTVL
           DO 10 I=1,NN
           H(I)=0.0
           V(I)=0.0
   10      IT(I)=0.0
           DUD=0.123456789
           FN=N
           FNN=NN
           FNNN=FNN*FNN
   50      NS=NS+1
           NG=IN
           HTS=0.0
   40      FK=NG
           HT=FK*(FNN-FK/(FNNN)
           HT=2.*HT
           HTS=HTS+HT
           H(NG)=H(NG)+HTS
           IT(NG)=IT(NG)+1
           IF(NG.LT.1) GO TO 20
           IF(NG.EQ.NN) GO TO 20
           FQ = FK/FNN
           NG=0
           DO 30 I=1,NN
           X=RANF(DUD)
           IF(X-FQ)31,30,30
   31      NG=NG+1
   30      CONTINUE
           GO TO 40
   20      IF(NS-IOUT)50,51,51
   51      DO 60 I=1,NN
           IF(IT(I))60,60,61
   61      V(I)=H(I)/FLOAT(IT(I))
   60      CONTINUE
           WRITE(6,70)NS,N, IN
   70      FORMAT(1H1,*RESULT OF SIMULATION*,5X,*NO. REPEATS=*,I4,/,
                                                        *POP SI
     1ZE =*,I3,5X,*INITIAL NO. MUT =*,I2)
           FQI=FLOAT(IN)/FNN
           DO 85 I=1,NN
           Y=FLOAT(I)/FNN
   85      TH(I)=8.*FN*T*(1./3.-Y/6.)-(4.0*FN*FQI*FQI)/3.0
           WRITE(6,75)
   75      FORMAT(1H0,*GENE FRQ =*,6X,*HET SIM =*,6X,*HET THR =*
           DO 80 I=2,NN
           FQQ=FLOAT(I)/FNN
   80      WRITE(6,81)FQQ,V(I),TH(I)
   81      FORMAT(1H ,F10.3,F15.5,F15.5)
           IOUT=IOUT+INTVL
           IF(NS-MAX)50,90,90
   90      STOP
```

END

RESULT OF SIMULATION   NO. REPEATS = 1800

POP SIZE = 10   INITIAL NO. MUT = 2

GENE FRQ =	HET SIM =	HET THR =
.100	1.92370	2.40000
.150	3.31096	3.56667
.200	4.35804	4.66667
.250	5.59055	5.70000
.300	6.67667	6.66667
.350	7.71280	7.56667
.400	8.12897	8.40000
.450	8.57295	9.16667
.500	9.65053	9.86667
.550	10.11152	10.50000
.600	10.64712	11.06667
.650	11.16056	11.56667
.700	11.96495	12.00000
.750	12.63243	12.36667
.800	13.45620	12.66667
.850	12.46155	12.90000
.900	12.24626	13.06667
.950	11.82384	13.16667
1.000	12.69651	13.20000

5.12 Random time change

In the preceding section, we selected certain paths from the original process by changing the probability measure with formula (5.4). Another class of transformations connected with additive functionals is the operation of "random time change".

Let $V(x)$ be an arbitrary, positive function. Define $\tau_t(\omega)$ by

$$(5.33) \qquad t = \int_0^{\tau_t(\omega)} V(x_{\xi,\omega})\,d\xi$$

where ω indicates a specific path and $x_{\xi,\omega}$ is the position of the sample path ω at time ξ. This is a path (ω) dependant integral. Let us regard $\tau_t(\omega)$ as a new time parameter. This time change means that at each point x the motion of the path is accelerated (or decelerated) with coefficient $V^{-1}(x)$. Under a random time change, the "road map" of a path remains exactly the same.

Now let $\tilde{p}(\tau, x, y)$ be the transition probability density that a path is at y at time τ measured by the new parameter, given that it is at x at $\tau = 0$. Then the infinitesimal operator of the process governed by $\tilde{p}(\tau, x, y)$ is

$$(5.34) \qquad \frac{1}{V(x)} A$$

where A is the operator for $p(t, x, y)$. Consider the random drift case with

$$A = \frac{x(1 - x)}{4N} \frac{\partial^2}{\partial x^2} .$$

Let

$$V(x) = 2x(1 - x)$$

which is equal to the heterozygote frequency. Then let, as in (5.33),

(5.35)
$$t = \int_0^{\tau_t(\omega)} 2x_{\xi\omega}(1 - x_{\xi\omega})d\xi .$$

The operator of the new process is

(5.36)
$$\frac{1}{V(x)} A = \frac{1}{2x(1 - x)} \frac{x(1 - x)}{4N} A = \frac{1}{8} \frac{d^2}{dx^2}$$

and therefore the new KBE is

(5.37)
$$\frac{\partial u}{\partial \tau} = \frac{1}{8} \frac{\partial^2 u}{dx^2} .$$

Note that the process governed by (5.37) is a Brownian motion which is the simplest diffusion process. This result asserts that the simple Brownian motion and the random drift process of genetics have exactly the same road map and that if a path of the latter process is accelerated (or decelerated) by the reciprocal of the heterozygosity at each locality, the process becomes a Brownian motion. In other words, if we measure time by the amount of heterozygosity which appeared in each path, the process becomes simple Brownian motion. By similar arguments, it can be shown that every one-dimensional diffusion process can be transformed into a Brownian motion. However it seems necessary to mention that not every random time change is useful, and that once a process is transformed according to a random time change, then there is no way of recovering the original process. The latter fact warns us that unless the function $V(x)$ of (5.33) is the quantity of our interest, or a quantity of interest can be obtained from it, the time change is not useful. In the above example (5.37), τ is the cummulative heterozygosity up to time t which is, of course, a familiar quantity. It should also be noted that the procedure used for the substitution of t for τ in (5.35) does not produce the same time transformation for any two paths.

CHAPTER 6

NUMERICAL INTEGRATION OF THE KOLMOGOROV BACKWARD EQUATION

6.1   Integration method

Consider the KBE

(6.1)
$$\frac{\partial u}{\partial t} = Au = \frac{V_{\delta x}}{2}\frac{\partial^2 u}{\partial x^2} + M_{\delta x}\frac{\partial u}{\partial x}$$

where

(6.2)
$$u = u(t, x) = \int_0^1 p(t, x, y)f(y)dy ,$$

p(t, x, y) is a transition probability density, and f(y) is an arbitrary function. The function u(t, x) is the expectation of the quantity f(y), as a function of x and t. By "integration" of the KBE
(6.1), we mean to solve (6.1) and obtain the solution which satisfies
the boundary conditions and the initial value u(0, x) = f(x). Here we
shall study a numerical method; the KBEs in population genetics problems are usually hard to solve explicitly. There are some cases which
can be solved in terms of eigenvalues and eigenvectors of the KBEs;
however, we will deal with this analytic method in the next chapter.
Here we seek an iterative solution to (6.1), and the subject is restricted to population genetics, though the analogy applies to any
diffusion process.

Note that u(t, x) is defined on a half-infinite strip $[0,\infty)\times[0,1]$,
i.e., $0 \leq t < \infty$ and $0 \leq x \leq 1$. Divide the space $[0,\infty)\times[0,1]$ into a
lattice, (see Fig. 6.1).

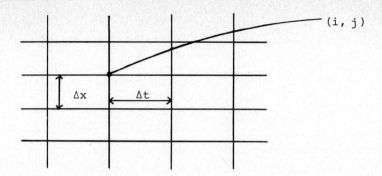

Fig. 6.1

We denote a grid point of the lattice by (i, j). Consider $u(t, x)$ over the grid points of the lattice, and denote by u_{ij} the value of $u(t, x)$ at (i, j), i.e. $u_{ij} = u(i, j)$. We approximate the derivatives by differences:

$$(6.3) \qquad \frac{\partial u(t, x)}{\partial t} \approx \frac{u_{i+1,j} - u_{ij}}{\Delta t} \, ,$$

$$(6.4) \qquad \frac{\partial u(t, x)}{\partial t} \approx \frac{u_{ij} - u_{i,j-1}}{\Delta x} \quad \text{or} \quad \approx \frac{u_{i,j+1} - u_{ij}}{\Delta x} \, ,$$

$$(6.5) \qquad \frac{\partial^2 u(t, x)}{\partial x^2} \approx \frac{u_{i,j+1} - 2u_{ij} + u_{i,j-1}}{(\Delta x)^2} \, .$$

Upon substitution of (6.3), (6.4) and (6.5) into (6.1), we replace the partial differential equation by the following difference equation

$$(6.6) \qquad \frac{u_{i+1,j} - u_{ij}}{\Delta t} = \frac{V_{\delta x_i}}{2} \frac{u_{i,j+1} - 2u_{ij} + u_{i,j-1}}{(\Delta x)^2}$$

$$+ M_{\delta x_i} \frac{u_{i,j+1} - u_{i,j-1}}{2\Delta x}$$

where $M_{\delta x_i}$ and $V_{\delta x_i}$ are understood to assume values of $M_{\delta x}$ and $V_{\delta x}$ at the corresponding lattice point. The difference equation (6.6) can be rearranged to

$$(6.7) \qquad u_{i+1,j} = u_{ij} + \frac{\Delta t V_{\delta x_i}}{2(\Delta x)^2} \left\{ u_{i,j+1} - 2u_{ij} + u_{i,j-1} \right\}$$

$$+ \frac{\Delta t M_{\delta x_i}}{2\Delta x} \left\{ u_{i,j+1} - u_{i,j-1} \right\}$$

where

(6.8)
$$u_{0j} = f(x_j).$$

Therefore starting from $i = 0$ ($t = 0$), we can obtain the values of u_{ij} successively for $i = 1, 2, 3, \cdots$.

In order to determine numerical values of u_{ij}, we need the boundary conditions (values) at $x = 0$ and $x = 1$. The general boundary conditions were discussed in chapter 4. We shall look at the following three types; absorbing and reflecting boundaries, as we considered before, and "adhesive" boundaries also of importance in genetics. At an adhesive boundary, a path remains in the boundary after having reached it. This is different from an absorbing boundary for which we disregard a path as soon as it reaches the boundary. The adhesive boundary is a special case of the return process, $p_{00} = 1$ or $p_{11} = 1$.

At the boundary $x = 0$, the boundary values are

(6.9)   absorbing: $u_{i0} = 0$

(6.10)   reflecting: $u_{i0} = u_{i1}$

                          (the value at the first lattice point)

(6.11)   adhesive: $u_{i0} = f(0)$

where u_{i0} corresponds to $u(t, 0) = f(0)$. The conditions are analogous at the boundary $x = 1$.

For the solution obtained by iteration (6.7) to be an approximation of the solution of the KBE (6.1), u_{ij} must converge to the true solution as Δx and Δt become small. But convergence is not guaranteed unless we choose the ratio of Δx to Δt properly. If we choose a wrong ratio, u_{ij} may diverge. A sufficient condition for the convergence in this case is

(6.12)
$$\max_{x_i} \frac{\Delta t V_{\delta x_i}}{2(\Delta x)^2} < 1$$

and

(6.13)
$$\max_{x_i} \frac{\Delta t M_{\delta x_i}}{2(\Delta x)} < 1.$$

If $M_{\delta x_i} = 0$ for all $i = 1, 2, \cdots$, (6.12) is the only requirement for the convergence.

Here we considered only time homogeneous processes and therefore $V_{\delta x}$ and $M_{\delta x}$ were independent of time. But the same method is applicable to time inhomogeneous process by substituting appropriate $V_{\delta x}(t)$ and $M_{\delta x}(t)$ which may vary in time.

6.2 Examples

Using the difference equation (6.7), the calculation of the mean gene frequency and the standard deviation for a hypothetical cage population of N = 500 were carried out. The selection model with dominance h and the selection coefficient s was used, (KBE 2.23):

Table 6.1 Mean and standard deviation
of the gene frequencies.

time (generations)	s = 0	s = 0.01			
		h = 0	h = 0.05	h = 0.5	h = 1.0
Mean frequency					
0	0.2	0.2	0.2	0.2	0.2
10	0.2	0.203	0.203	0.208	0.213
50	0.2	0.218	0.220	0.242	0.265
100	0.2	0.240	0.244	0.286	0.330
Standard deviation					
10	0.0399	0.0418	0.0441	0.0409	0.0397
50	0.088	0.0959	0.0969	0.0963	0.0956
100	0.123	0.146	0.147	0.147	0.142

The following is the calculation for the standard deviation of the gene frequency for the following frequency dependent selection model:

genotype	AA	Aa	aa
fitness	$1 + s(x - \frac{1}{2})$	1	$1 + s(\frac{1}{2} - x)$

Table 6.2 Standard deviation of the gene frequency.

time (generation) \ s	0	0.01	0.03	0.10
10	0.0499	0.0488	0.0467	0.0404
50	0.1104	0.0992	0.0814	0.0502
100	0.1543	0.1270	0.0911	0.0504

(For all cases, x_0 = 0.5 and N = 500).

Note that for small s, the standard deviation increases as t increases, but for a large s, the deviation nearly stabilizes at early generations (compare the columns under s = 0 and s = 0.1).

For the pure random drift case (2.5), the mean gene frequency, the average heterozygosity, and the probability of fixation were calculated by the numerical method, and they are presented in the Fig. 6.1 ～ 6.3. With u(0, x) = f(x) = x, u(t, x) is the mean gene frequency at time t, given that it starts from x. With f(x) = 2x(1-x), u(t, x) is the average heterozygosity, and f(x) = 0 for x < 1 and f(1) = 1, u(t, x) is the probability that a path reaches fixation by time t.

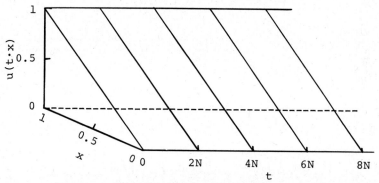

Fig. 6.1 The mean gene frequency u(t, x) as the function of time (t generations) and initial frequency (x). u(t, x) is the solution of (2.5) and (6.1) with f(x) = x.

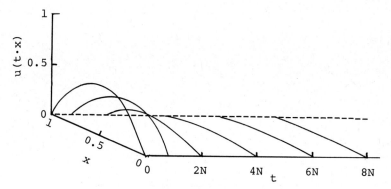

Fig. 6.2 The mean heterozygosity u(t, x) as the function of t and x. U(t, x) is the solution of (2.5) and (6.1) with f(x) = 2x(1-x).

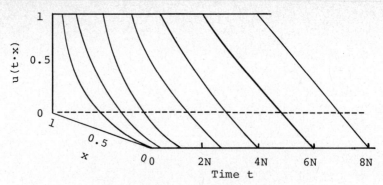

Fig. 6.3 The probability of fixation up
to time t as the function of x.
u(t, x) is the solution of (2.5) and
(6.1) with f(x) = 0 if x < 1 and
f(x) = 1 if x = 1.

The following is a listing of a computer program which calculates
the solution of the KBE according to the scheme given by (6.7).

```
C       THIS IS A PROGRAM TO INTEGRATE KOLMOGOROV BACKWARD EQUATION OF
C       POPULATION GENETICS
        DIMENSION B(100),U(100),F(100),V(100),E(100)
        DIMENSION SS(10),HS(10)
        SS(1)=0.0
        SS(2)=0.01
        SS(3)=0.03
        SS(4)=0.1
        SS(5)=0.3
        SS(6)=0.5
        SS(7)=1.0
        HS(1)=0.0
        HS(2)=0.05
        HS(3)=0.5
        HS(4)=0.95
        HS(5)=1.0
        M=20
        N=500
        NST=5
        NDF=5
        DO 180 IBD=1,NST
        DO 180 IAD=1,NDF
        TL=100.0
        TINT=10.0
        FM=M
        FN=N
        MM=M+1
        S=SS(IBD)
        H=HS(IAD)
        WRITE (6,190)N,S,H
190     FORMAT(3H1N=,I5,10X,2HS=,F6.3,10X,2HH=,F6.3)
C       VARIANCE AND MEAN COEFFICIENTS OF EQUATION
        DO 10 I=2,M
        X=FLOAT(I-1)/FM
```

```
       Y=1.-X
       E(I)=(2.0*S*X*Y*(0.5-X)/(1.0=2.0*S*(0.5-X)**2)
   10  V(I)-(X*Y)/(2.*FN)
C      DETERMINE DELTA TIME
       T=1.0
       DO 20 I=2,M
       T1=1.0   /(V(I)*FM*FM)
       IF(E(I)-0.00001)57,57,58
   57  T2=1.0
       GO TO 59
   58  T2=2.0/(E(I)*FM)
   59  IF(T1-T)21,22,22
   21  T=T1
   22  IF(T2-T)23,20,20
   23  T=T2
   20  CONTINUE
       DT=T
C      INITIAL CONDITION
       DO 30 I=1,MM
       X=FLOAT(I-1)/FM
   30  F(I)=X
       DO 40 I=2,M
       V(I)=(DT*V(I)*FM*FM)/2.0
   40  E(I)=DT*E(I)*FM/2.0
       TIME=0.
       DO 41 I=1,MM
   41  U(I)=F(I)
   50  TOUT=TIME+TINT
   51  TIME=TIME+DT
       B(1)=F(1)
       B(MM)=f(MM)
       DO 60 I=2,M
   60  B(I)=U(I)+V(I)*(U(I+1)+U(I-1)-2.0*U(I))+E(I)*(U(I+1)-U(I-1))
       DO 61 I=1,MM
   61  U(I)=B(I)
       IF(TIME-TOUT)51,62,62
   62  WRITE(6,80) TIME
       WRITE(6,81)(U(I),I=1,MM)
       IF(TIME-TL)50,180,180
   80  FORMAT(6H0TIME=,F10.2)
   81  FORMAT(10F10.6)
  180  CONTINUE
   70  STOP
       END
```

CHAPTER 7

EIGENVALUES AND EIGENVECTORS OF THE KBE

7.1 Eigenvalues and eigenvectors

Consider the KBE

(7.1) $$\frac{\partial u}{\partial t} = Au$$

where

(7.2) $$u = u(t, x) = \int_0^1 p(t, x, y) f(y) dy.$$

As before, consider (7.2) as a linear transformation on a space of functions into itself. Namely

(7.3) $$u(t, x) = T_t f(x).$$

Then (7.1) is

(7.4) $$\frac{\partial T_t f(x)}{\partial t} = AT_t f(x) = T_t Af(x)$$

(see (1.62) for the comutativity of T_t and A, i.e., $T_t A = AT_t$). The operator T_t is a linear operator with a parameter t on a function space. The linearity of T_t means that

$$T_t(af(x) + g(x)) = \int_0^1 p(t, x, y)\{af(y) + g(y)\}dy$$

$$= a\int_0^1 p(t, x, y)f(y)dy + \int_0^1 p(t, x, y)g(y)dy$$

$$= aT_t f + T_t g , \qquad \text{where } a \text{ is a constant.}$$

Now suppose that $e(x)$ is a function such that

(7.5)
$$Ae(x) + \lambda e(x) = 0$$

where A is the operator in (7.1) and (7.4), and λ is a scaler.   Then
the operator T_t has a special property that

(7.6)
$$\frac{\partial T_t e(x)}{\partial t} = T_t Ae(x) = -\lambda T_t e(x).$$

Hence if we let $u(t,x) \equiv T_t e(x)$, then equation (7.6) can be written as

$$\frac{du(t,x)}{dt} = -\lambda u(t,\ x).$$

Therefore we have

$$u(t,x) = u(0,x) e^{-\lambda t}$$

which is

(7.7)
$$T_t e(x) = e(x) e^{-\lambda t}$$

$(T_0 e(x) = e(x))$.

Hence if a function (or a quantity) satisfies the relationship (7.5),
the expectation of the quantity at time t is given simply by the exponential formula (7.7).

A function which satisfies (7.5) and which is not everywhere zero
is called an eigenfunction of the operator A or T_t, and the number λ
associated with an eigenfunction is called an eigenvalue.

We have shown that if a quantity happens to be given by an eigenfunction, the expectation at some future time can be obtained by the
simple formula (7.7).   There would be a similar situation if a quantity could be expressed as a linear combination of eigenfunctions.
Suppose $\{e_k(x)\}$ and $\{\lambda_k\}$ are the eigenfunctions and eigenvalues, respectively, i.e.,

(7.8)
$$Ae_k(x) + \lambda_k e_k(x) = 0 \ .$$

If $f(x)$ can be expressed as a linear combination of $e_k(x)$, i.e.,

(7.9)
$$f(x) = \sum_k a_k e_k(x)$$

then

(7.10)
$$T_t f(x) = T_t \sum_k a_k e_k(x) = \sum_k a_k T_t e_k(x) = \sum_k a_k e_k(x) e^{-\lambda_k t}.$$

This is the expectation of $f(x)$ at time t, and the formula is simple. Therefore if we obtain sufficiently many eigenfunctions so that any function can be expressed as a linear combination of these eigenfunctions, then the expectation of any function can be obtained by formula (7.10). Then the question is how many eigenfunctions are sufficient? Usually it requires a large number of linearly independent eigenfunctions, and almost always, infinitely many. In cases of population genetics the differential operator A in (7.1) has usually countably many eigenfunctions and their associated eigenvalues are discrete and distributed on the half open interval $[0, \infty)$. Furthermore, every square-integrable function can be expressed as a linear combination of the eigenfunctions. That is, if $f(x) \in L^2[0, 1]$, then there is a series

$$(7.11) \qquad\qquad \Sigma\, a_k e_k(x)$$

such that

$$(7.12) \qquad \lim_{n \to \infty} \int_0^1 |f(x) - \sum_k^n a_k e_k(x)|^2 dx = 0.$$

We cannot, however, generally expect a stronger convergence, such as

$$(7.13) \qquad \text{Max}_x\ |f(x) - \sum_k^n a_k e_k(x)| \to 0$$

as $n \to \infty$. Convergence (7.12) is called "convergence in the mean of order 2" and (7.13) is known as "uniform convergence" which is the strongest kind.

Good references for the eigenfunction expansion are E. A. Coddington & N. Levinson (1955), Chap. 7, 9, 11 and 12; E. C. Titchmarsh, (1946) vol. 1; Courant and Hilbert (1962).

It is now clear that one way to solve the KBE is to find a complete set of eigenfunctions and eigenvalues of the operator A. Then we express $f(x)$ as a linear combination of the eigenfunctions. Another property which is more important is that the first few dominant eigenvalues and the associated eigenfunctions reveal a lot of information about the process.

To find the characteristic pairs (eigenvalues and eigenfunctions), we need to solve the differential equation (7.5) which can be rewritten as

$$(7.14) \qquad\qquad [A + \lambda]e(x) = 0$$

in which λ and $e(x)$ are a pair of unknowns.

The difficulty of equation (7.14) in cases of population genetics is that the operator A does not have simple solutions. This is because the coefficient $V_{\delta x}$ of the operator A usually vanishes at x = 0 and x = 1.

In general, when the coefficient of the highest order in a differential equation vanishes at a point, say p_0 (singularity point), a solution assuming a finite value at another point, say p_1 (regular point), diverges as the argument of the solution approaches p_0. The radius of convergence of a power series expansion of the solution at p_1 is equal to $|p_0 - p_1|$.

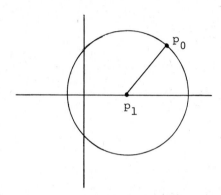

Fig. 7.1   Radius of convergence of the power series $\sum_i a_i (z - p_1)^i$ on the complex plane.

When there are any points at which the leading coefficient vanishes, a differential equation admits at most one linearly independent solution which is bounded at both of the two singular points, whereas if the leading coefficient never vanishes a second order equation has two linearly independent solutions which satisfy preassigned boundary conditions.

7.2   Pure random drift case

Let us examine in detail the situation of the simplest case

(7.15)
$$A = \frac{x(1 - x)}{4N} \frac{d^2}{dx^2} .$$

This example and the next one are simple and can be found in a mathematical table. However we will discuss them here in detail for two purposes: to show how characteristic pairs are obtained and used, and

to show that essentially the same principle can be applied to more
complicated situations which cannot be found in a mathematical table.

With A in (7.15), equation (7.14) becomes

$$\frac{x(1-x)}{4N}\,e''(x) + \lambda e(x) = 0 .$$

We can rewrite this as

(7.16) $$x(1-x)e''(x) + \Lambda e(x) = 0$$

in which $\Lambda = 4N\lambda$. It is natural to express the bounded solution of
(7.15) as a power series

(7.17) $$e(x) = \sum_{i=0}^{\infty} a_i x^i .$$

Note that we do not allow i to be a negative integer, though generally
a solution expanded at a singular point has some or infinitely many
negative powers. Differentiating e(x) in (7.17) twice,

(7.18) $$e''(x) = \sum_i i(i-1)a_i x^{i-2} .$$

Substituting this into (7.15), we have

(7.19) $$\sum_i i(i-1)a_i x^{i-1} - \sum_i i(i-1)a_i x^i + \Lambda \sum_i a_i x^i = 0 .$$

This implies

(7.20) $$i(i+1)a_{i+1} - i(i-1)a_i + \Lambda a_i = 0 \qquad \text{for all i .}$$

Hence

(7.21) $$a_{i+1} = \frac{i(i-1) - \Lambda}{i(i+1)}\, a_i .$$

This recurrence relation determines all the coefficients (a_i) in the
power series expansion of e(x). Since e(x) in (7.17) is expanded
about x = 0 and operator A has another singularity at x = 1, the e(x)
will diverge unless the power series terminates with finite number of
terms and it terminates only if $i(i-1)-\Lambda = 0$. We can list the values
of Λ for which the series terminates:

Λ	i	a_1	a_2	a_3	a_4	a_5
0	1	1				
2	2	1,	-1			
6	3	1,	-3,	2		
12	4	1,	-6,	10,	-5	
20	5	1,	-10,	30,	-35,	14
.	.	.	.	.	.	.
.	.	.	.	.	.	.
.	.	.	.	.	.	.

The coefficients in the above table are calculated as follows:
From (7.21), if Λ is equal to one of the integers $i(i-1)$ the series
terminates at a_i, i.e., $a_{i+1} = 0$. For fixed value of Λ, we can calculate a_i using (7.21). For example, $\Lambda = 2\cdot3 = 6$ (i = 3), $a_1 = 1$,

$$a_2 = \frac{2\cdot0 - 6}{1\cdot2}\, a_1 = -\frac{6}{2}\, a_1 = -3 \ ,$$

$$a_3 = \frac{2\cdot1 - 6}{2\cdot3}\, a_2 = \frac{-4}{6}(-3) = \frac{12}{6} = 2$$

and

$$a_4 = \frac{2\cdot3 - 6}{3\cdot4}\, a_3 = 0 \ .$$

The first few characteristic pairs calculated in this way are

λ_k	$e_k(x)$
0	$e_1(x) = x$
$1/(2N)$	$e_2(x) = x - x^2$
$3/(2N)$	$e_3(x) = x - 3x^2 + 2x^3$
$6/(2N)$	$e_4(x) = x - 6x^2 + 10x^3 - 5x^4$
$10/(2N)$	$e_5(x) = x - 10x^2 + 30x^3 - 35x^4 + 15x^5$
.	.
.	.
.	.

To illustrate the use of these characteristic pairs, we will calculate some quantities relevant to genetics. For example, the second
moment of the gene frequency is given by

$$f(x) = x^2 = e_1(x) - e_2(x) .$$

Hence the second moment at time t is precisely

$$u(t, x) = T_t f(x) = e_1(x) e^{0t} - e_2(x) e^{-\frac{t}{2N}}$$

$$= x - x(1 - x)^{-\frac{t}{2N}}$$

$$= x \left\{ 1 - (1 - x) e^{-\frac{1}{2N}} \right\} .$$

The variance of the gene frequency is equal to the second moment minus the square of the mean, which is

$$x \left\{ 1 - (1 - x) e^{-\frac{t}{2N}} \right\} - x^2 = x(1 - x) \left\{ 1 - e^{-\frac{t}{2N}} \right\} .$$

(Note that mean = $T_t x = x e^{0t} = x$.)

Let us have a slightly more complicated example, the variance of the heterozygosity. To do this, we need to obtain its second moment. Note that the relavent function can be expressed as

$$f(x) = \{2x(1 - x)\}^2 = 4x^2 - 8x^3 + 4x^4$$

$$= \frac{4}{5} \{e_2(x) - e_4(x)\} .$$

Hence the second moment of the heterozygosity is

$$T_t f(x) = \frac{4}{5} \left\{ e_2(x) e^{-\frac{t}{2N}} - e_4(x) e^{-\frac{6t}{2N}} \right\}$$

$$= \frac{4}{5} \left\{ x(1 - x) e^{-\frac{t}{2N}} - x(1 - x)(1 - 5x + 5x^2) e^{-\frac{6t}{2N}} \right\}$$

$$= \frac{4}{5} x(1 - x) \left\{ e^{-\frac{t}{2N}} - (1 - 5x + 5x^2) e^{-\frac{6t}{2N}} \right\} .$$

Noting that the mean of the heterozygosity is

$$2x(1 - x) e^{-\frac{t}{2N}} ,$$

the variance is therefore

$$\sigma_h^2 = \frac{4}{5} x(1-x)\left\{ e^{-\frac{t}{2N}} - (1 - 5x + 5x^2)e^{-\frac{6t}{2N}} \right\} - 4x^2(1-x)^2 e^{-\frac{2t}{2N}}$$

$$= 4x(1-x)\left\{ \frac{1}{5} e^{-\frac{t}{2N}} - \frac{1 - 5x + 5x^2}{5} e^{-\frac{6t}{2N}} - x(1-x) e^{-\frac{2t}{2N}} \right\}.$$

Note that $\sigma_h^2 = 0$ if $t = 0$ or $t = \infty$, as it ought to be. When does σ_h^2 become maximum?

The system of eigenfunctions constructed in this section is known as ultraspherical polynomials or Gegenbauer polynomials, which is a special case of more general polynomials known as Jacobi polynomials. But usually these polynomials are dealt on a slightly different domain. Substitution of $y = 1-2x$ in equation (7.16) yields

$$(1 - y^2)e''(y) + \Lambda e(y) = 0$$

which is now defined on an interval $[-1, 1]$ instead of $[0, 1]$. For this equation, eigenfunctions are given by

$$e_0(y) = 1 ,$$

$$e_1(y) = -y$$

and

$$e_{n+1}(y) = \frac{1}{n+1}\left\{ (2n-1)ye_n(y) - (n-2)e_{n-1}(y) \right\} \qquad \text{for } n \geq 1 .$$

And the associated eigenvalues are

$$\Lambda_n = n(n - 1) .$$

Kimura (1955) applied the ultraspherical polynomials to solve the pure random drift problem. General references are Erdélyi (1953), chapter X or Szegö (1967), chapter IV, and also Abramowitz and Stegun (1964), chapter 22, or any reference book which deals with classical orthogonal polynomials. Different authors may use different notations and different normalizations.

7.3 Mutation and random drift case

Next consider the model with mutation, discussed in section 2.5. We shall first examine the case of reversible mutation, $v > 0$ and $u > 0$. Then

(8.22) $$A = \frac{x(1-x)}{4N}\frac{d^2}{dx^2} + \{v - (u + v)x\}\frac{d}{dx}$$

The differential equation (7.14) in this case is

$$\frac{x(1 - x)}{4N} e''(x) + \{v - (u + v)x\}e'(x) + \lambda e(x) = 0$$

which is equivalent to

(7.23) $x(1 - x)e''(x) + \{V - (U + V)x\}e'(x) + \Lambda e(x) = 0$

where $V = 4Nv$, $U = 4Nu$ and $\Lambda = 4N\lambda$.
As before let

$$e(x) = \Sigma\, a_i x^i,$$

$$e'(x) = \Sigma\, i a_i x^{i-1},$$

$$e''(x) = \Sigma\, i(i - 1)a_i x^{i-2}.$$

Substituting these formulae into (7.23), we have that

$$\sum_{i=0}^{\infty}\left\{i(i-1)a_i x^{i-1} i(i-1)a_i x^i + Vi a_i x^{i-1} - (U+V)i a_i x^i + \Lambda a_i x^i\right\} = 0 \ .$$

So

$$(i+1)i a_{i+1} x^i - i(i-1)a_i x^i + V(i+1)a_{i+1} x^i$$

$$- (U+V)(i+1)a_{i+1} x^i + \Lambda a_i x^i = 0$$

and thus

$$a_{i+1} = \frac{i(i - 1) + i(U + V) - \Lambda}{i(i + 1) + V(i + 1)}\, a_i \ .$$

Hence a_i will terminate if Λ is equal to $i(i-1)+i(U+V)$, for $i = 0, 1, 2, \cdots$, and the first few pairs are

Λ	i	a_0	a_1	a_2
0,	0,	1		
U+V,	1,	1,	$-(U+V)/V$	
2(1+U+V),	2,	1,	$-2(1+U+V)/V$,	$(1+U+V)(2+U+V)/V(1+V)$
3(2+U+V),	3,	1,	$-3(2+U+V)/V$,	$3(3+U+V)(2+U+V)/V(1+V)$
$\vdots$	$\vdots$	$\vdots$		

$$a_3 = \frac{(2+U+V)(3+U+V)(4+U+V)}{V(1+V)(2+V)}$$

Therefore

$$\lambda_0 = 0, \qquad\qquad e_0(x) = 1 ;$$

$$\lambda_1 = U+V, \qquad\qquad e_1(x) = 1 - \frac{U+V}{V}\,x ;$$

$$\lambda_2 = \frac{2(1+U+V)}{4N}, \qquad e_2(x) = 1 - \frac{2(1+U+V)}{V}\,x + \frac{(1+U+V)(2+U+V)}{V(1+V)}\,x^2 ;$$

$$\lambda_3 = \frac{3(2+U+V)}{4N}, \qquad e_3(x) = 1 - \frac{-3(2+U+V)}{V}\,x + \frac{3(3+U+V)(2+U+V)}{V(1+V)}\,x^2$$
$$\qquad\qquad - \frac{(2+U+V)(3+U+V)(4+U+V)}{V(1+V)(2+V)}\,x^3 ;$$

$$\lambda_4 = \frac{4(3+U+V)}{4N}, \qquad e_4(x) = 1 - \frac{4(3+U+V)}{V}\,x + \frac{6(3+U+V)(4+U+V)}{V(1+V)}\,x^2$$
$$\qquad\qquad - \frac{4(3+U+V)(4+U+V)(5+U+V)}{V(1+V)(2+V)}\,x^3$$
$$\qquad\qquad + \frac{(3+U+V)(4+U+V)(5+U+V)(6+U+V)}{V(1+V)(2+V)(3+V)}\,x^4 .$$

The mean gene frequency is $f(x) = x = \frac{V}{v+u}\left\{e_0(x) - e_1(x)\right\}$. Hence

$$(7.24) \qquad T_t f(x) = \frac{V}{u+v}\,e_0(x) - \frac{V}{u+v}\,e_1(x)\,e^{-(u+v)t}$$

$$= \frac{V}{u+v} - \left(\frac{V}{u+v} - x\right)e^{-(u+v)t} .$$

This means that if a path starts from x, it goes eventually to $v/(u+v)$ and the approaching rate is $(u+v)$ which is the sum of forward and backward mutation rates, (see Crow and Kimura, 1970, p. 390). The second moment can be calculated with

$$f(x) = x^2 = \frac{v(1+v)}{(1+v+u)(2+v+u)}\,e_2(x) - \frac{2v(1+v)}{(u+v)(2+u+v)}\,e_1(x)$$
$$+ \frac{v(1+v)}{(u+v)(1+u+v)}\,e_0(x).$$

Hence

$$T_t f(x) = \frac{v(1+v)}{(1+u+v)(2+u+v)}\,e_2(x)\,e^{-\frac{2(1+u+v)t}{4N}}$$

$$- \frac{2v(1+v)}{(u+v)(2+u+v)}\,e_1(x)\,e^{-(u+v)t} + \frac{v(1+v)}{(u+v)(1+u+v)}$$

(see Crow and Kimura p. 391).
The second moment eventually reduces to

$$\frac{v(1+v)}{(u+v)(1+u+v)}$$

and the variance of the gene frequency is thus

$$\frac{v(1+v)}{(u+v)(1+u+v)} - \left(\frac{v}{u+v}\right)^2 = \frac{uv}{(1+u+v)(u+v)(u+v)} = \frac{\bar{x}(1-\bar{x})}{1+4N(u+v)}$$

where $\bar{x} = \dfrac{v}{u+v}$ (see Crow and Kimura P. 440, 9. 2. 10).

7.4 Irreversible mutation case

Setting $v = 0$ in the case given by (7.22) corresponds to the model of infinite alleles (or irreversible mutation). Although it is a special case of (7.22), it requires a different treatment.

The differential equation to be solved for the characteristic pairs is

(7.25) $x(1-x)e''(x) - Uxe'(x) + \Lambda e(x) = 0$

where $U = 4Nu$ and $\Lambda = 4N\lambda$, corresponding to (7.23). As before let

$$e(x) = \sum_i a_i x^i.$$

Substituting this and its derivatives into (7.25), we obtain

(7.26) $a_{i+1} = \dfrac{i(i-1) + iU - \Lambda}{i(i+1)} a_i.$

The recurrence relation (7.26) terminates for $\Lambda = U$, $2(1+U)$, $3(2+U)$, $4(3+U)$, $\cdots$. The first few pairs are

$e_k(x)$	λ_k
$e_1(x) = x$	u
$e_2(x) = x - \dfrac{2+U}{2} x^2$	$2(1+U)/4N$
$e_3(x) = x - \dfrac{2(3+U)}{2} x^2 + \dfrac{(3+U)(4+U)}{2\cdot 3} x^3$	$3(2+U)/4N$
$e_4(x) = x - \dfrac{3(4+U)}{2} x^2 + \dfrac{3(4+U)(5+U)}{2\cdot 3} x^3$ $\qquad - \dfrac{(4+U)(5+U)(6+U)}{2\cdot 3\cdot 4} x^4$	$4(3+U)/4N$

Note that the unity function $e_0(x) \equiv 1$ and $\Lambda = 0$ are no longer a characteristic pair. The mean gene frequency is given by $f(x) = x = e_1(x)$ and therefore

$$T_t e_1(x) = e_1(x) e^{-ut} = x e^{-ut}.$$

The mean frequency decreases exponentially with the mutation rate, u, but it is independent of the population size.

7.5 Hypergeometric differential equation

Having examined the pure random drift model and the reversible and irreversible mutation models, we should note that there exists a simple rule to obtain the characteristic pairs.

The differential equation which the pairs must satisfy in these cases can be written as

(7.27) $\quad x(1-x)e''(x) + \{c - (a+b+1)x\}e'(x) - abe(x) = 0$

where a, b and c are scalers. This is called the hypergeometric differential equation. If we let

$$e(x) = \Sigma\, a_i x^i$$

and substitute this in (7.27), we have

(7.28) $\qquad a_{i+1} = \dfrac{(i+a)(i+b)}{(i+1)(i+c)}\, a_i.$

Therefore the solution is, if $c > 0$

(7.29) $\quad e(x) = 1 + \dfrac{a \cdot b}{1 \cdot c} x + \dfrac{a \cdot (a+1)b(b+1)}{1 \cdot 2 \cdot c \cdot (c+1)} x^2$

$$+ \dfrac{a \cdot (a+1)(a+2) \cdot b \cdot (b+1)(b+2)}{1 \cdot 2 \cdot 3 \cdot c \cdot (c+1)(c+2)} x^3 + \cdots$$

and if $c = 0$

(7.30) $\quad e(x) = x + \dfrac{(a+1)(b+1)}{1 \cdot 2 \cdot 1} x^2$

$$+ \dfrac{(a+1)(a+2)(b+1)(b+2)}{1 \cdot 2 \cdot 3 \cdot 1 \cdot 2} x^3 + \cdots .$$

Obviously $ab = -\Lambda$ and for $b = 0, -1, -2, -3, \cdots$ the expansion of $e(x)$ in (7.29) terminates, and for $b = -1, -2, -3, \cdots$ expansion of $e(x)$ in (7.30) terminates.

In the case of the pure random drift model, $c = 0$ and $a+b+1 = 0$. Thus

(7.31) $a = -(1 + b)$,

(7.32) $\Lambda = b(1 + b)$.

With substitution of $b = -1$ into (7.29), we obtain the first eigenfunction, and the associated eigenvalue $\Lambda = 0$. And with $b = -2$, the
second eigenfunction and the eigenvalue $\Lambda = 2$ which implies $\lambda = 1/2N$
since $\Lambda = 4N\lambda$, and so on. In the case of the reversible mutation
model, $c = V$ and $a+b+1 = U+V$, so that formula (7.29) applies to this
case. Also

$$\Lambda = -b(U + V - b - 1).$$

Letting $b = 0$ in (7.29), we obtain the first eigenfunction associated
with $\lambda_0 = 0$, and with $b = -1$ in (7.29), we have the second eigenfunc-
tion associated with $\Lambda_1 = U+V$ or $\lambda_1 = u+v$ since $\Lambda_1 = 4N\lambda_1$; with $b = -2$,
we have the third eigenfunction associated with $\lambda_2 = 2(U+V+1)/4N$, and
so on.

   Formulae (7.28), (7.29) and (7.30) are simple, We may ask: do
these formulae apply to all the cases of population genetics? The
answer is unfortunately "no".

   The type of differential equations to which essentially the same
formulae as (7.28) - (7.30) apply is of the form

(7.33) $f''(x) + \left\{ \dfrac{A}{x} + \dfrac{B}{x - 1} \right\} f'(x)$

$$+ \left\{ \dfrac{C}{x^2} - \dfrac{D}{x(1 - x)} + \dfrac{E}{(x - 1)^2} \right\} f(x) = 0.$$

(This is not however the most general form.) Note that the coefficient
of $f'(x)$ in (7.33) has poles of order 1, whereas the coefficient of
$f''(x)$ has poles of order 2. The differential equation in (7.33) can
be transformed to the hypergeometric equation given in (7.27).
References: Ahlfors (1953); Whittaker and G. N. Wattson (1927); Ince
(1934).

7.6 Orthogonality of eigenfunctions

   We will next examine the orthogonality of eigenfunctions. The
eigenfunctions are usually orthogonal with respect to a certain weight
function. Namely, there is a positive function $w(x)$ such that

$$\int_0^1 w(x) e_k(x) e_\ell(x) \, dx = 0 \qquad \text{if } k \neq \ell$$

The theory of differential equations asserts that if the differential equation is of the form

$$A e_k(x) + \lambda_k e_k(x) = 0$$

which is

(7.34)
$$\frac{V_{\delta x}}{2} e_k''(x) + M_{\delta x} e'_k(x) + \lambda_k e_k(x) = 0,$$

the weight function is

(7.35)
$$w(x) = \frac{1}{V_{\delta x}} e^{\displaystyle 2 \int_0^x \frac{M_{\delta \xi}}{V_{\delta \xi}} \, d\xi}.$$

This view of looking at a system of eigenfunctions associated with an operator gives a unified view to the eigenfunction expansion of functions. The system of eigenfunctions is therefore a system of orthogonal functions in the function space $L^2[0, 1; w(x)]$ where $w(x)$ is given by (7.35). In the terminology of functional analysis, the space $L^2[0, 1; w(x)]$ is called a "Hilbert space", and operator A is called a "linear operator" on the Hilbert space $L^2[0, 1; w(x)]$.

7.7 Expansion by eigenfunctions

Let us first introduce some notation. For f, g $\in L^2[0, 1; w]$, we write

$$\int_0^1 w(x) f(x) g(x) \, dx = \langle f, g \rangle$$

and

$$\int_0^1 w(x) f(x) f(x) \, dx = \langle f, f \rangle = \|f\|^2.$$

The orthogonality has an important consequence when we write an arbitrary function in terms of the eigenfunctions. Let $f(x) \in L^2[0, 1; w]$, and use the familiar expansion:

(7.36)
$$f(x) = \sum_i a_i e_i(x).$$

How do we determine the coefficients in (7.36)? Very simple! If we multiply both sides by $e_j(x)w(x)$ and integrate, we have

$$\int_0^1 e_j(x)w(x)f(x)dx = \sum_i a_i \int_0^1 e_j(x)w(x)e_i(x)dx .$$

Noting the orthogonality $<e_j, e_i> = 0$ if $i \neq j$, we have

$$<f, e_j> = \int_0^1 e_j(x)w(x)f(x)dx = a_j \|e_j\|^2$$

and hence

(7.37)
$$a_j = \frac{<f, e_j>}{\|e_j\|^2} .$$

Therefore

(7.38)
$$f(x) = \sum_i \frac{<f, e_i>}{\|e_i\|^2} e_i(x).$$

This is called a "generalized Fourier series". A very good reference on this subject is Kolmogorov and Fomin (1961).

If we apply T_t from (7.3) on both sides of (7.38), we have

(7.39)
$$T_t f(x) = \sum_i \frac{<f, e_i>}{\|e_i\|^2} e_i(x) e^{-\lambda_i t} .$$

This is the solution of the KBE.

As a simple consequence of formula (7.39), we have

$$\lim_{t \to \infty} T_t f(x) = \frac{<f, e_0>}{\|e_0\|^2} e_0(x) = \frac{<f, 1>}{\|1\|^2}$$

if $e_0(x) = 1$ and $\lambda_0 = 0$ belong to $L^2[0, 1; w]$. A case of special importance is

$$f(x) = \delta(x - y),$$

where

$$T_t f(x) = T_t \delta(x - y)$$

is the probability density that a path is at y given that it starts from x. Therefore, if $\lambda_0 = 0$ and $e_0(x) = 1$ are the characteristic pairs, $T_t \delta(x - y)$ eventually approaches an equilibrium function $\phi(y)$ which is given by

$$\frac{<\delta(x - y), 1>}{\|1\|^2}$$

where

$$<\delta(x - y), 1> = \int_0^1 w(x)\delta(x - y)dx = w(y)$$

$$= \frac{1}{V_{\delta y}} e^{2\int_0^y \frac{M_{\delta\xi}}{V_{\delta\xi}} d\xi}$$

and

$$\|1\|^2 = \int_0^1 w(x)dx .$$

Therefore the equilibrium distribution is

$$(7.40) \qquad \phi(x) = \frac{1}{\|1\|^2 V_{\delta x}} e^{2\int_0^x \frac{M_{\delta\xi}}{V_{\delta\xi}}} = \frac{w(x)}{\|1\|^2} .$$

This is "Wright's formula" in the general form. For the case of reversible mutation discussed above,

$$w(x) = x^{V-1}(1 - x)^{U-1}$$

and

$$\|1\|^2 = \|e_0\|^2 = B(V, U) .$$

Therefore, in this case, the equilibrium formula is

$$(7.41) \quad \phi(x) = \frac{x^{V-1}(1 - x)^{U-1}}{B(V, U)} = \frac{\Gamma(4N(U+V))}{\Gamma(4Nu)\Gamma(4Nv)} x^{4Nv-1}(1 - x)^{4Nu-1}$$

while most authors choose to use the Kolmogorov forward equation for such results, all the quantities derived here have been obtained from the KBE.

From formula (7.10) we can see why the first few eigenfunctions and eigenvalues are important. If it has 0 as an eigenvalue, then $T_t f(x)$ approaches the equilibrium at the rate of the next eigenvalue. If 0 is not an eigenvalue, then it will not have an equilibrium, but everything will asymptotically vanish at the rate of the first eigenvalue.

References are S. Wright (1969); Crow and Kimura (1970); Ewens (1969); Moran (1962).

7.8 The steady-state distribution of gene frequencies

Some explicit forms of Wright's formula are given below. The gene frequency distribution in a finite population of size N with immigrants from a population with a constant gene frequency $(\bar{x})$ is given by

$$(7.42) \qquad \phi(x) = \frac{x^{4Nm\bar{x}-1}(1 - x)^{4Nm(1-\bar{x})-1}}{B(4Nm\bar{x}, \ 4Nm(1-\bar{x}))}$$

where m is the migration rate. This formula is equivalent to (7.41): If we let $v = m\bar{x}$ and $u = m(1 - \bar{x})$ this reduces to (7.41).

The distribution for the genic selection model with reversible mutation given by the KBE

$$(7.43) \qquad \frac{\partial u}{\partial t} = \frac{x(1 - x)}{4N}\frac{\partial^2 u}{\partial x^2} + [sx(1 - x) + v - (u + v)x]\frac{\partial u}{\partial x}$$

is

$$(7.44) \qquad \phi(x) = Ce^{4Nsx}x^{4Nv-1}(1 - x)^{4Nu-1}$$

where C is a constant so that $\int_0^1 \phi(x)dx = 1$. The case of a completely recessive favorable allele is

$$(7.45) \qquad \phi(x) = Ce^{2Nsx^2}x^{4Nv-1}(1 - x)^{4Nu-1}$$

where s is the selective advantage of the recessive allele. In the more general case, if the relative fitnesses of AA, Aa and aa are s, sh and 0, respectively, measured in Malthusian parameters, then the frequency distribution of allele A is given by

$$(7.46) \qquad \phi(x) = Ce^{2Nsx^2+4Nshx(1-x)}x^{4Nv-1}(1 - x)^{4Nu-1}.$$

It is evident that a special case of (7.46) with h = 0 reduces to (7.45). The following figure compares these distributions when v = u. Population size (N) ranges from $1/(40u)$, $10/(40u)$ and $100/(40u)$, and selection also varies from a value almost negligible to a higher order.

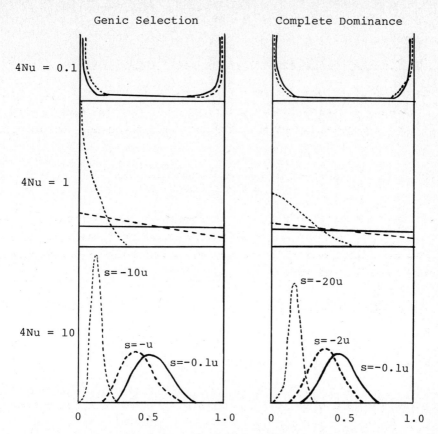

Fig. 7.1 Equilibrium distributions of gene
frequencies. From Wright (1969, p. 365).

The distribution of a completely recessive lethal gene is

$$(7.47) \qquad \phi(x) = C(1 - x^2)^{2N} x^{4Nv-1} (1 - x)^{-1}$$

where v is the mutation rate from a normal gene to the lethal. The
backward mutation from the lethal to normal is ignored because the
lethal genes are rare. The frequency distribution of completely recessive lethals is shown in the following figure for $v = 10^{-5}$ and
various population sizes.

See Crow and Kimura (1970), pp. 445-449, and Nei (1968) for par-
tially dominant lethal genes and the distribution of lethal carrying
chromosomes. For the distribution of self-sterility alleles in finite
populations, see Wright (1939, 1965).

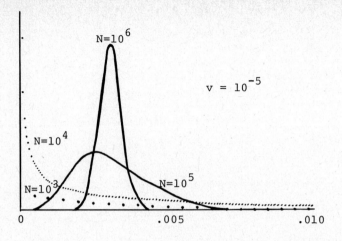

Fig. 7.2 Equilibrium distribution of lethal
   recessive genes. From Wright (1969, p. 366).

CHAPTER 8

APPROXIMATION METHODS

8.1 Perturbation

We shall mainly discuss perturbation methods which are often useful, particularly when a specific problem deviates only slightly from
an analytically solvable case.

Suppose that an equation

(8.1) $Ae(x) + \lambda e(x) = 0$

can be solved. Let the characteristic pairs be $\{e_k(x), \lambda_k\}$. Now let
us change (8.1) slightly and consider

(8.2) $Ae(x) + \{\lambda + \varepsilon q(x)\}e(x) = 0$

where ε is a small number. We call this a perturbed equation.

We assume that the eigenvalues $\{\tilde{\lambda}_k\}$ and eigenfunction $\{\tilde{e}_k(x)\}$ of
the perturbed equation (8.2) are analytic functions of ε, so there are
expansions of the form

(8.3) $\tilde{\lambda}_k = \lambda_k + \varepsilon\lambda_k^{(1)} + \varepsilon^2\lambda_k^{(2)} + \cdots$

(8.4) $\tilde{e}_k(x) = e_k(x) + \varepsilon e_k^{(1)}(x) + \varepsilon^2 e_k^{(2)}(x) + \cdots$

where $\lambda_k^{(i)}$ are constants and $e_k^{(i)}(x)$ are unknown functions.

Suppose that we can operate upon these with the perturbed operator, and equate the coefficients of ε^k in the resulting formulae.
Suppose also that the function $e_k^{(1)}$, $e_k^{(2)}$, $\cdots$ occurring in (8.4) can
be expanded in terms of the unperturbed eigenfunctions in the form

(8.5)
$$e_k^{(1)}(x) = \sum_i a_{ki}^{(1)} e_i(x),$$

(8.6)
$$e_k^{(2)}(x) = \sum_i a_{ki}^{(2)} e_i(x)$$

etc. This is in fact possible because $\{e_i(x)\}$ forms a complete dense set or a basis in the function space we are dealing with.

The values of the coefficients in (8.3) and (8.4) can be found as follows. The perturbed eigenfunctions satisfy

(8.7)
$$A\tilde{e}_k(x) + (\tilde{\lambda}_k + \varepsilon q(x))\tilde{e}_k(x) = 0.$$

Hence by substituting (8.3) and (8.4) into (8.7) we have

(8.8)
$$Ae_k + \varepsilon Ae_k^{(1)} + \varepsilon Ae_k^{(2)} + \cdots$$
$$+ \{\lambda_k + \varepsilon \lambda_k^{(1)} + \varepsilon^2 \lambda_k^{(2)} + \cdots + \varepsilon q\}\{e_k + \varepsilon e_k^{(1)} + \cdots\} = 0.$$

Equating the coefficients of ε to 0, we obtain

(8.9)
$$Ae_k + \lambda_k e_k = 0,$$

(8.10)
$$Ae_k^{(1)} + \lambda_k e_k^{(1)} + (\lambda_k^{(1)} + q)e_k = 0,$$

(8.11)
$$Ae_k^{(2)} + \lambda_k e_k^{(2)} + [\lambda_k^{(1)} + q]e_k^{(1)} + \lambda_k^{(2)} e_k = 0$$

etc. Substituting (8.5) into (8.10), we have

(8.12)
$$A(\sum_i a_{ki}^{(1)} e_i) + \lambda_k (\sum_i a_{ki}^{(1)} e_i) + (\lambda_k^{(1)} + q)e_k = 0.$$

Noting that

$$A(\sum_i a_{ki}^{(1)} e_i) = - \sum_i a_{ki}^{(1)} \lambda_i e_i ,$$

(8.12) becomes

(8.13)
$$\sum_i a_{ki}^{(1)} \lambda_i e_i - \lambda_k (\sum_i a_{ki}^{(1)} e_i) - (\lambda_k^{(1)} + q)e_k = 0 .$$

On multiplying this equation by $w(x)e_k(x)$ and integrating over $(0,1)$, we have

$$a_{kk}^{(1)}\lambda_k\|e_k\|^2 - \lambda_k a_{kk}^{(1)}\|e_k\|^2 - \lambda_k^{(1)}\|e_k\|^2 - \langle e_k, qe_k\rangle = 0$$

therefore

(8.14)
$$\lambda_k^{(1)} = \frac{-\langle e_k, qe_k\rangle}{\|e_k\|^2}$$

where e_k is the eigenfunction of the unperturbed equation (8.1), and

$$\langle e_k, qe_k\rangle = \int_0^1 e_k^2(x)q(x)w(x)dx.$$

Similarly on multiplying (8.13) by $w(x)e_\ell(x)$ and integrating over $(0, 1)$, we obtain

$$a_{k\ell}^{(1)}\lambda_\ell\|e_\ell\|^2 - \lambda_k a_{k\ell}^{(1)}\|e_\ell\|^2 - \langle e_\ell, qe_k\rangle = 0 .$$

Hence

(8.15)
$$a_{k\ell}^{(1)} = \frac{-\langle e_\ell, qe_k\rangle}{(\lambda_k - \lambda_\ell)\|e_\ell\|^2} , \qquad k \neq \ell.$$

Observe that the eigenfunctions of the perturbed equation are orthogonal with respect to the same weight function, w, as the unperturbed one, i.e.,

$$\langle \tilde{e}_k, \tilde{e}_\ell\rangle = \int_0^1 \tilde{e}_k\tilde{e}_\ell w\,dx = 0 \qquad \text{if } k \neq \ell .$$

Since eigenfunctions can be normalized by a constant, assume that

$$\|e_i\|^2 = \|\tilde{e}_i\|^2 = 1, \qquad i = 1, 2, \cdots.$$

Then

$$1 = \|e_k\|^2 = \|e_k + \varepsilon e_k^{(1)} + \varepsilon^2 e_k^{(2)} + \cdots\|^2$$

$$= \|e_k\|^2 + 2\varepsilon\langle e_k, e_k^{(1)}\rangle + \varepsilon^2 \cdots.$$

This implies $\langle e_k, e_k^{(1)}\rangle = 0$, and therefore

$$0 = <e_k, \sum_i a_{ki}^{(1)} e_i> = a_{kk}^{(1)} \|e_k\|^2.$$

Thus

$$(8.16) \qquad a_{kk}^{(1)} = 0.$$

With (8.14), (8.15) and (8.16), we have determined the first coefficients of (8.3) and (8.4). Namely

$$\lambda_k^{(1)} \text{ of } (8.3) = \frac{-<e_k, qe_k>}{\|e_k\|^2}$$

$$e_k^{(1)}(x) \text{ of } (8.4) = -\sum_{i \neq k} \frac{<e_i, qe_k>}{(\lambda_k - \lambda_i)\|e_i\|^2} e_i.$$

The higher coefficients of (8.3) and (8.4) can be obtained similarly, but it requires more calculations. The first order approximation of the purturbed eigenfunction and the second order approximation of the eigenvalue are respectively

$$(8.17) \qquad \tilde{e}_k(x) = e_k(x) - \varepsilon \sum_{i \neq k} \frac{<e_i, qe_k>}{(\lambda_k - \lambda_i)\|e_i\|^2} e_i(x), \text{ and}$$

$$(8.18) \qquad \tilde{\lambda}_k = \lambda_k - \varepsilon \frac{<e_k, qe_k>}{\|e_k\|^2} - \varepsilon^2 \sum_{i \neq k} \frac{<e_i, qe_k>^2}{(\lambda_k - \lambda_i)\|e_i\|^2}.$$

For details, see Chapt. 7, T. Kato (1966).

8.2 Examples

The selection model given by (2.22) can be considered as a perturbation of the pure random drift case. The differential equation for the characteristic pairs is

$$(8.19) \qquad x(1-x)f'' + 4Nsx(1-x)f' + \Lambda f = 0,$$

while the equation for the random drift case is

$$(8.20) \qquad x(1-x)f'' + \Lambda f = 0.$$

Equation (8.19) is not quite the form of perturbation given in (8.2).

In order to reduce (8.19) to the form of (8.2), we set $f(x) = g(x)e^{-Sx}$ where $S = 2Ns$.   Then

$$f' = -Se^{-Sx}g + e^{-Sx}g \; ,$$

$$f'' = S^2e^{-Sx}g - 2Se^{-Sx}g' + e^{-Sx}g'' \; .$$

On substituting these into (8.19), we have

(8.21) $\quad\quad\quad x(1 - x)g'' + [\Lambda - S^2x(1 - x)]g = 0$

in which $-S^2$ and $x(1-x)$ should be considered as ε and $q(x)$ in (8.2) respectively.   Recall $e_1 = x(1-x)$ and $w(x) = 1/x(1-x)$.   Hence

$$\langle e_1, qe_1 \rangle = \int_0^1 \frac{[x(1 - x)]^2 x(1 - x)}{x(1 - x)} = \frac{1}{30}$$

and

$$\langle e_1, e_1 \rangle = \|e_1\|^2 = \frac{1}{6} \; ,$$

so that

$$\frac{\langle e_1, qe_1 \rangle}{\|e_1\|^2} = \frac{1}{5} \; .$$

Therefore, if S is small, the first order approximation of the first eigenvalue is

$$\tilde{\Lambda}_1 \approx \Lambda_1 + \frac{S^2}{5} \; .$$

To obtain the second eigenvalue,

$$\langle e_2, e_2 \rangle = \frac{1}{120}$$

and

$$\langle e_2, qe_2 \rangle = \frac{2}{105} \; ,$$

so that

$$\frac{\langle e_2, qe_2 \rangle}{\|e_2\|^2} = \frac{16}{7} \; .$$

Therefore

$$\tilde{\Lambda}_2 \approx \Lambda_2 + \frac{16}{7} S^2 \; ,$$

and thus

$$\tilde{\lambda}_1 = \frac{\Lambda_1}{4N} \approx \frac{1}{2N} + \frac{N}{5} s^2 ,$$

$$\tilde{\lambda}_2 = \frac{\Lambda_2}{4N} \approx \frac{3}{2N} + \frac{16N}{7} s^2 .$$

(see Crow and Kimura (1970) p. 398 for similar case with a higher order approximation).

Consider the following model of overdominance,

Genotype	AA	Aa	aa
Fitness measured in Malthusian parameter	$-s_1$	0	$-s_2$

Then the KBE for this case is

$$\frac{\partial u}{\partial t} = \frac{x(1 - x)}{4N} \frac{\partial^2 u}{\partial x^2} + [s_2 - (s_1 + s_2)x]x(1 - x)\frac{\partial u}{\partial x} .$$

The differential equation which the characteristic pairs must satisfy is

$$x(1 - x)f" + 4N[s_2 - (s_1 + s_2)x]x(1 - x)f' + \Lambda f = 0$$

where $\Lambda = 4N\lambda$. This equation can be rewritten as

$$f" + 4N[s_2 - (s_1 + s_2)x]f' + \frac{\Lambda}{x(1 - x)} f = 0 .$$

Miller (1962) has worked out this problem using another method. First he observes that as $4Ns_1$ and $4N(s_1 + s_2)$ become large, the eigenfunction can be approximated by a normal function, i.e.,

$$e(x) \sim e^{-(x - \hat{x})^2}$$

where

$$\hat{x} = \frac{s_2}{s_1 + s_2} = \text{the equilibrium value as Ns} \to \infty .$$

Then, using a method of integral equation, he obtains a formula for the dominant eigenvalue with large Ns_1 and Ns_2 values. He has found an interesting fact that as the equilibrium value $\hat{x}$ tends to 0 or to

1, the dominant eigenvalue increases as Ns increases. The value of
the dominant eigenvalue decreases as Ns increases if $\hat{x}$ is not near 0
or 1, because the overdominance effect is to retard the probability
of reaching fixation or extinction. Table 8.1 shows the relationship
between the values of λ_1 for given $\hat{x}$ and $c = N(s_1 + s_2)/2$.

Table 8.1 Values of $2N\lambda_1$ as a function of
$c = N(s_1+s_2)/2$ and $\hat{x}$. (From Miller, 1962).

$\hat{x}$ $\quad$ c	0.5	0.6	0.7	0.8	0.9
0	1	1	1	1	1
0.5	.813	.817	.829	.849	.846
1.0	.653	.668	.713	.787	.892
1.5	.517	.548	.641	.793	1.004
3.0	.236	.315	.552	.935	1.453
5.0	.068	.157	.455	1.011	1.837
10.0	.001	.019	.192	.858	2.262

8.3 Numerical method

We should also know that to find the dominant eigenvalue and its
associated eigenfunction for a given model (or KBE) we can use the
numerical integration method dealt with in chapter 6. For example, to
find the first characteristic pair for an operator with exit boundaries
at $x = 0$ and $x = 1$, use $f(x) = x(1-x)$ and integrate. Then for sufficiently large t,

$$\frac{\int_0^1 u(t,\ x)dx - \int_0^1 u(t+1,\ x)dx}{\int_0^1 u(t,\ x)dx}$$

will converge to the dominant eigenvalue with $u(t,\ x)$ as a function of
x as the associated eigenfunction. Symbolically,

$$\lambda = \lim_{t\to\infty} \frac{\int_0^1 u(t,\ x)dx - \int_0^1 u(t+1,\ x)dx}{\int_0^1 u(t,\ x)dx}.$$

This is a simple, but a good method. The method gives us both eigenvalue and eigenfunction.

8.4 Singular perturbation

It is sometimes said that the diffusion model is valid only when
the deterministic pressure is weak, but it is not otherwise. This is
a wrong assertion and, in fact, as the deterministic pressure becomes
much stronger than the stochastic force, the solution of the KBE converges to that of the entirely deterministic process for each fixed x.
The outline of the reason is as follows.

Consider the general KBE

$$(8.22) \qquad \frac{\partial u}{\partial t} = \frac{V_{\delta x}}{2} \frac{\partial^2 u}{\partial x^2} + M_{\delta x} \frac{\partial u}{\partial x}$$

and assume that

$$(8.23) \qquad \frac{V_{\delta x}}{2} << |M_{\delta x}| \qquad \text{for all } x .$$

This means that the deterministic force is much stronger than the
stochastic force. Let us assume that we can rewrite $\frac{V_{\delta x}}{2}$ as

$$(8.24) \qquad \varepsilon \cdot \frac{\hat{V}_{\delta x}}{2} .$$

which means that $V_{\delta x}$ can be made uniformly small. Then the KBE is

$$(8.25) \qquad \frac{\partial u}{\partial t} = \varepsilon \cdot \frac{\hat{V}_{\delta x}}{2} \frac{\partial^2 u}{\partial x^2} + M_{\delta x} \frac{\partial u}{\partial x} .$$

This KBE can be considered as a perturbation of the deterministic
equation

$$(8.26) \qquad \frac{\partial u}{\partial t} = M_{\delta x} \frac{\partial u}{\partial x} .$$

The view that equation (8.25) is a perturbed equation of (8.26)
has one difficulty in that this perturbation converts a first order equation (8.26) to the second order equation (8.25). This is called
"singular perturbation", and it does not have a converging series.
Instead, we must be content with an asymptotic perturbation which is
valid only for $\varepsilon \to 0$.

The theory of perturbation asserts that if $U_0(t, x)$ is the solution of the deterministic case (8.26), then the solution of the

perturbed equation is

(8.27) $\quad u(t, x) = U_0(t, x) + \varepsilon \int_0^t \int_0^1 P_0(t-\xi, x, z) A U_0(\xi, z) dz d\xi$

where

$$A = \frac{\hat{V}_{\delta x}}{2} \frac{\partial^2}{\partial x^2} \quad \text{of (8.25)}$$

and $P_0(t, x, y)$ is the transition probability of the unperturbed equation. In this case $P_0(t, x, y)$ is degenerate because the unperturbed process is deterministic.

From formula (8.27), it is clear that as ε becomes small the solution converges to the deterministic solution. The convergence is for each fixed x, but it is not necessarily "uniformly" convergent. That is, for an arbitrarily small ε we may find a range of x in which formula (8.27) is not a good approximation, or in other words, if we denote the solution given in (8.27) by $U_\varepsilon(t, x)$,

$$\lim_{\varepsilon \to 0} \lim_{x \to 0} U_\varepsilon(t, x) \underset{\substack{\text{does not} \\ \text{necessarily equal}}}{=} \lim_{x \to 0} \lim_{\varepsilon \to 0} U_\varepsilon(t, x).$$

At any rate the solution of the KBE in which $M_{\delta x}$ dominates $V_{\delta x}$ should be as valid as the solution where $V_{\delta x}$ is about equal to or dominates the deterministic force $M_{\delta x}$. Reference: Wasow (1965).

CHAPTER 9

GEOGRAPHICAL STRUCTURE OF POPULATIONS

9.1 One-dimensional populations, discrete colonies

Consider a population consisting of n colonies occupying a circular habitat. Migration occurs between adjacent colonies at a rate m.
Each colony size is N, every mutant is unique, the mutation rate is u,
and all alleles are selectively neutral. The circularity of the population is for mathematical convenience, and analogous results hold for
a linear case with two ends.

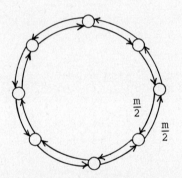

Fig. 9.1 Circular model; small circles
indicate colonies and arrows indicate
migration.

Unlike a panmictic population, with this or the model shown later, it
is difficult to formulate a Markov process of gene frequency change.
However there is a quantity that can be easily studied for this type
of population model. That quantity is the covariance of the gene frequencies among colonies, which is a "second order process".

Let f_{ti} be the probability that at time t two randomly chosen homologous genes, separated by a distance of i colonies, are "identical by descent", for i = 0, 1, 2, $\cdots$ n-1, and $f_{t,n} = f_{t,0}$.

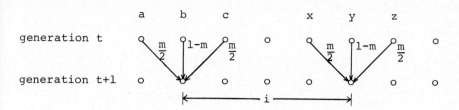

$$
\begin{array}{ccccccc}
& a & b & c & x & y & z
\end{array}
$$

Fig. 9.2 Diagram illustrating paths of
genes from generation t to t+1.

Then $f_{t,i}$ satisfies the following recurrence equation:

$$(9.1) \quad f_{t+1,i} = (1-u)^2 \left[\underset{(a,\ x)}{(\tfrac{m}{2})^2 f_{t,i}} + \underset{(a,\ y)}{\tfrac{m(1-m)}{2} f_{t,i+1}} + \underset{(a,\ z)}{(\tfrac{m}{2})^2 f_{t,i+2}} \right.$$

$$+ \underset{(b,\ x)}{\tfrac{(1-m)m}{2} f_{t,i-1}} + \underset{(b,\ y)}{(1-m)^2 f_{t,i}} + \underset{(b,\ z)}{\tfrac{(1-m)m}{2} f_{t,i+1}}$$

$$\left. + \underset{(c,\ x)}{(\tfrac{m}{2})^2 f_{t,i-2}} + \underset{(c,\ y)}{\tfrac{m(1-m)}{2} f_{t,i-1}} + \underset{(c,\ z)}{(\tfrac{m}{2})^2 f_{t,i}} \right]$$

in which the letters in the parentheses indicate the paths diagramed in Fig. 9.2, and $(1-u)^2$ is the probability that two genes under consideration have not mutated. Rearranging the terms according to the subscript i, we have

$$(9.2) \quad f_{t+1,i} = (1-u)^2 \left[(\tfrac{m}{2})^2 f_{t,i-2} + m(1-m) f_{t,i-1} \right.$$

$$\left. + \left\{ (1-m)^2 + \tfrac{m^2}{2} \right\} f_{t,i} + m(1-m) f_{t,i+1} + (\tfrac{m}{2})^2 f_{t,i+2} \right].$$

This recurrence equation can be expressed neatly in a matrix notation.
Let

$$F_t \equiv \begin{bmatrix} f_{t,0} \\ f_{t,1} \\ \cdot \\ \cdot \\ \cdot \\ f_{t,n-1} \end{bmatrix}$$

(9.3)

$$R \equiv \begin{pmatrix} 0 & 1 & & & & \\ & 0 & 1 & & & \\ & & \cdot & \cdot & & \\ & & & \cdot & \cdot & \\ & & & & \cdot & 1 \\ 1 & & & & & 0 \end{pmatrix} \quad \text{all unspecified elements are 0}$$

$$R^{-1} = \begin{pmatrix} 0 & & & & & 1 \\ 1 & 0 & & & & \\ & 1 & \cdot & & & \\ & & \cdot & \cdot & & \\ & & & \cdot & \cdot & \\ & & & & 1 & 0 \end{pmatrix}$$

$$R^2 = \begin{pmatrix} 0 & 0 & 1 & & & \\ & 0 & 0 & 1 & & \\ & \cdot & \cdot & \cdot & & \\ & & \cdot & \cdot & \cdot & \\ & & & \cdot & \cdot & 1 \\ 1 & & & \cdot & \cdot & 0 \\ 0 & 1 & & & & 0 \end{pmatrix}$$

$$R^{-2} = \begin{pmatrix} 0 & 0 & & & 1 & 0 \\ 0 & 0 & \cdot & & & 1 \\ 0 & 1 & \cdot & \cdot & & \\ & & \cdot & \cdot & \cdot & \\ & & & \cdot & \cdot & 0 & 0 \\ & & & 1 & & \end{pmatrix}.$$

Using these matrices and noting $f_{t,n} = f_{t,0}$, equation (9.2) can be rewritten as

(9.4)
$$F_{t+1} = (1 - u)^2 \left[(\tfrac{m}{2})^2 R^{-2} + m(1 - m)R^{-1} \right.$$
$$\left. + \left\{ (1 - m)^2 + \tfrac{m^2}{2} \right\} I + m(1 - m)R + (\tfrac{m}{2})^2 R^2 \right] F_t.$$

One modification is necessary in the above equation. When two genes in generation t come from a single colony in the previous generation, it is possible that they are descendants of a single gene, and this probability is $1/2N$. Therefore if two genes come from a single colony, instead of $f_{t,0}$ we need to replace it by

(9.5)
$$\left(1 - \tfrac{1}{2N} \right) f_{t,0} + \tfrac{1}{2N} = f_{t,0} + \frac{1 - f_{t,0}}{2N} .$$

Let M be the matric in the square brackets in (9.4), which is a polynomial in R, and let

(9.6)
$$g_t \equiv \begin{pmatrix} \dfrac{1 - f_{t,0}}{2N} \\ 0 \\ 0 \\ \vdots \\ 0 \end{pmatrix} .$$

Then, the correct recurrence relation is

(9.7)
$$F_{t+1} = (1 - u)^2 M(F_t + g_t) .$$

The existence and uniqueness of the solution for recurrence equation (9.7) are shown in Appendex II.

Using this recurrence equation, we can obtain the vector F_t for any initial condition F_0.

At equilibrium, $F_{t+1} = F_t$ and therefore

$$F = (1 - u)^2 M(F + g)$$

or

(9.8)
$$\left\{ I - (1 - u)^2 M \right\} F = (1 - u)^2 Mg .$$

This equation can be solved explicitly in terms of eigenfunctions or in this case eigenvectors and eigenvalues. Let

$$(9.9) \qquad e_k \equiv \begin{pmatrix} 1 \\ \cos \dfrac{2\pi k}{n} \\ \cos \dfrac{2\pi 2k}{n} \\ \cdot \\ \cdot \\ \cdot \\ \cos \dfrac{2\pi(n-1)k}{n} \end{pmatrix} \quad , \quad k = 0, 1, 2, \cdots, \left[\dfrac{n+1}{2}\right]$$

where $\left[\dfrac{n+1}{2}\right]$ = n/2 if n = even, = (n+1)/2 if n = odd. Note that

$$(9.10) \qquad (R + R^{-1})e_k = \cos \frac{2\pi k}{n} e_k = \xi_k e_k$$

where R is given in (9.3) and $\xi_k = \cos \dfrac{2\pi k}{n}$. Therefore e_k is an eigen-vector of R, and also of M, i.e.,

$$(9.11) \qquad M_e k = \lambda_k e_k$$

where

$$(9.12) \qquad \lambda_k = \left\{ 1 - m(1 - \cos \frac{2\pi k}{n}) \right\}^2 .$$

Since $\{e_k\}$ forms a basis of the vector space $(x_1, x_2, \cdots, x_n)$ where $x_1 = x_n$, $x_2 = x_{n-1}$, etc, we can expand any vector in this space by the vectors in this basis. Since F belongs to this space,

$$(9.13) \qquad F = \sum_{k=0}^{\left[\frac{n+1}{2}\right]} a_k e_k .$$

How do we determine the coefficients, a_k? Note that

$$(9.14) \qquad \langle e_k, e_\ell \rangle = \sum_{i=0}^{n} \cos \frac{2\pi ik}{n} \cos \frac{2\pi i\ell}{n}$$

$$= 0 \qquad \text{if } k \neq \ell$$

$$= n \qquad \text{if } k = \ell = 0$$

$$\qquad\qquad \text{or } k = \ell = \frac{n}{2}$$

$$= \frac{n}{2} \qquad \text{otherwise.}$$

On substituting (9.13) in (9.8) and noting that e_k's are eigenvectors of M, we obtain

$$(9.15) \qquad \sum_{k=0}^{\left[\frac{n+1}{2}\right]} \{1 - (1 - u)^2 \lambda_k\} a_k e_k = (1 - u)^2 Mg.$$

Also note that the matrix M is symmetric, and therefore for any two vectors x and y,

$$(9.16) \qquad \langle x, My \rangle = \langle Mx, y \rangle.$$

Noting (9.14) and (9.16), if we take the inner product of both sides of (9.15) with respect to eigenvector e_ℓ, all terms vanish, except

$$
\begin{aligned}
\{1 - (1 - u)^2 \lambda_k\} a_\ell \langle e_\ell, e_\ell \rangle &= \langle e_\ell, (1 - u)^2 Mg \rangle \\
&= (1 - u)^2 \langle Me_\ell, g \rangle \\
&= (1 - u)^2 \lambda_\ell \langle e_\ell, g \rangle \\
&= \frac{(1 - u)^2 \lambda_\ell (1 - f_0)}{2N},
\end{aligned}
$$

(see (9.6) for g). Therefore

$$(9.17) \qquad a_k = \frac{(1 - u)^2 (1 - f_0) \lambda_k}{2N \|e_k\|^2 \{1 - (1 - u)^2 \lambda_k\}}$$

where $\|e_k\|^2 = \langle e_k, e_k \rangle$ and λ_k is given in (9.12). The explicit formula is therefore

$$(9.18) \qquad F = \frac{(1-u)^2 (1-f_0)}{2N} \sum_{k=0}^{\left[\frac{n+1}{2}\right]} \frac{\lambda_k}{\|e_k\|^2 \{1 - (1-u)^2 \lambda_k\}} e_k$$

or

$$(9.19) \qquad f_i = \frac{(1-u)^2 (1-f_0)}{2N} \sum_{k=0}^{\left[\frac{n+1}{2}\right]} \frac{\lambda_k}{\|e_k\|^2 \{1 - (1-u)^2 \lambda_k\}} \cos \frac{2\pi i k}{n}.$$

In formulae (9.18) and (9.19), f_0 is implicit, but it can be determined explicitly,

$$f_0 = \frac{(1 - u)^2 w}{2N + (1 - u)^2 w}$$

where

$$w = \sum_{k=0}^{\left[\frac{n+1}{2}\right]} \frac{\lambda_k}{[1 - (1 - u)^2 \lambda_k]} \; .$$

An interesting relationship follows from formula (9.19). Let $\bar{f}$ be the average of f_i, that is the probability that two randomly chosen genes from the whole population are identical by descent. Then

$$\bar{f} = \frac{1}{n}\sum_{i=0}^{n-1} f_i = \frac{(1-f_0)(1-u)^2}{2Nn^2} \sum_{i=1}^{n-1} \sum_{k=0}^{\left[\frac{n+1}{2}\right]} \frac{k}{\|e_k\|^2[1-(1-u)^2\lambda_k]} \cos\frac{2\pi ik}{n} \; .$$

On interchanging the order of the summations, and noting

$$\sum_{i=0}^{n-1} \cos\frac{2\pi ik}{n} = 0 \qquad \text{if } k \neq 0$$

we have

(9.20)
$$\bar{f} = \frac{(1 - f_0)(1 - u)^2}{2Nn\{1 - (1 - u)^2\}}$$

$$\approx \frac{(1 - f_0)(1 - 2u)}{4Nnu} \approx \frac{1 - f_0}{4Nnu} \; .$$

In the last formula, higher order terms of u are neglected. If we let $nN = N_T$, the above formula is

$$\bar{f} = \frac{1 - f_0}{4N_T u} \; .$$

Noting that

$$\frac{(1 - u)^2(1 - f_0)}{2N} \frac{\lambda_k}{\{1 - (1 - u)^2\lambda_k\}} > 0$$

for all K, let us denote this quantity by w_k^2. Then formula (9.18) becomes

(9.21)
$$F = \sum_{k=0}^{\left[\frac{n+1}{2}\right]} \frac{w_k^2}{\|e_k\|^2} e_k \; .$$

This is a form often used for a second order process. The set of vectors $\{e_k\}$ is called the spectrum and $\left\{ w_k^2 / \|e_k\|^2 \right\}$ is called the spectral density. In this case, the spectral density is descrete and finite, (Maruyama, 1970).

Now let n be very large. Then the summations in formulae (9.18) and (9.19) can be approximated by integrals,

$$(9.22) \qquad f_i = \frac{(1-f_0)(1-u)^2}{2N\pi} \int_0^\pi \frac{\{1-m(1-\cos\theta)\}^2}{1-(1-u)^2\{1-m(1-\cos\theta)\}^2} \cos i\theta \, d\theta \ .$$

Analogous to (9.21), this can be written as

$$(9.23) \qquad f_i = \int_0^\pi w^2(\theta)\cos i\theta \, d\theta \ .$$

Here the spectral density is continuous, but located within a finite interval $[0, \pi]$. (Formula 9.22 was obtained by Kimura and Weiss, 1964.)

9.2   Continuous space

A continuous space analogue of the above model is as follows. The population is   continuous (circle) in space of length L, and the population density (D) is uniform. The total population size is LD = N_T. Let f(t, x) (the analogue of f_{ti}) be the probability that two genes separated by distance x are identical by descent, and r(x) be the probability density that two genes change their distance by x in one generation. If m(x) is the migration density of one gene, then

$$r(x) = \int m(\xi)m(x - \xi)d\xi \ .$$

The equation analogous to (9.2) is

$$f(t+1, x) = (1-u)^2 \int_{-\infty}^\infty r(\xi)f(t, x-\xi)d\xi$$

in which f(t, x) is a periodic function of x with periodicity L. As above, the correction for two genes coming from zero distance is necessary. The correction is

$$\frac{1 - f(t, 0)}{2D} \quad (x|\bmod L)$$

where x|mod L means that x is reduced by module of L, i.e. x|mod L = x - nL < L for some integern n. The corrected equation is therefore

(9.24)
$$f(t+1, x) = (1 - u)^2 \int_{-\infty}^{\infty} r(\xi) \left\{ f(t, x-\xi) \right.$$

$$\left. + \frac{1 - f(t, 0)}{2D} \delta(x-\xi \,|\, \text{mod } L) \right\} d\xi .$$

This recurrence equation provides $f(t, x)$ for any given initial condition, (Malécot, 1948, 1967, 1975; Maruyama 1971).

For the equilibrium, $f(t+1, x) = f(t, x)$. A standard procedure is to expand $f(x)$ in a Fourier series of the form

$$f(x) = \sum_{k=0}^{\infty} a_k \cos \frac{2\pi k x}{L} .$$

Upon substituting this in the equation with $f(t, x) = f(t+1, x)$, we can determine the coefficients. In doing this we need to use the orthogonality

$$\langle \cos \frac{2\pi k x}{L}, \cos \frac{2\pi \ell x}{L} \rangle = \int_0^L \cos \frac{2\pi k x}{L} \cos \frac{2\pi \ell x}{L} dx$$

$$= 0 \qquad \text{if } k \neq \ell ,$$

$$= L \qquad \text{if } k = \ell = 0, \text{ and}$$

$$= L/2 \qquad \text{if } k = \ell \neq 0 .$$

The equiliblium formula is

(9.25)
$$f(x) = \frac{(1-f_0)(1-u)^2}{2DL} \sum_{k=0}^{\infty} \frac{\Delta_k r_k}{\left\{1 - (1-u)^2 r_k\right\}} \cos \frac{2\pi k x}{L}$$

where

$$r_k = \int_0^{\infty} r(\xi) \cos \frac{2\pi k \xi}{L} d\xi$$

$\Delta_0 = 1$ and $\Delta_k = 2$ for $k \neq 0$. For example, if

$$r(x) = \frac{1}{\sqrt{4\pi\sigma^2}} e^{-\frac{x^2}{4\sigma^2}} \qquad \text{(normal distribution)},$$

$$r_k = e^{-\dfrac{4\pi^2 k^2 \sigma^2}{L^2}} \ .$$

From (9.25) we can obtain

$$\bar{f} = \frac{1}{L}\int_0^L f(x)\,dx = \frac{(1-f(0))(1-u)^2}{2DL(1-(1-u)^2)} \approx \frac{(1-f_0)(1-u)^2}{4N_Tu} \approx \frac{1-f_0}{4N_Tu}$$

which is equivalent to (9.20).

To make (9.25) analogous to (9.21) and (9.23) we rewrite

(9.26)
$$f(x) = \sum_{k=0}^{\infty} w_k^2 \cos\frac{2\pi kx}{L} \ .$$

As $L \to \infty$

(9.27)
$$f(x) = \int_0^{\infty} w^2(\theta)\cos\theta x\,d\theta \ .$$

The $f(x)$ in (9.26) has a spectral density which is discrete but infinite. The spectral density for (9.27) is continuous and distributed in the infinite interval $[0, \infty)$. Therefore we have encountered four different kinds of spectrums.

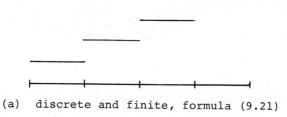

(a) discrete and finite, formula (9.21)

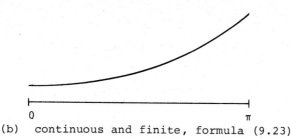

0   π
(b) continuous and finite, formula (9.23)

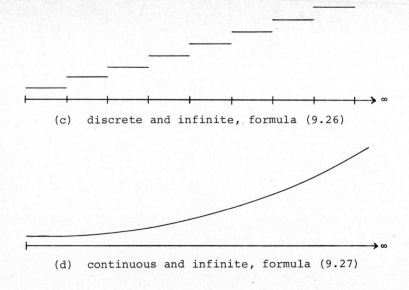

(c) discrete and infinite, formula (9.26)

(d) continuous and infinite, formula (9.27)

Fig. 9.3 Diagrams of spectral densities.

9.3 Two-dimensional populations

Essentially the same method used for the one dimensional case can be applied to a two-dimensional problem. The simplest case is to consider the product of the circular lattices shown in the diagram below. The two circular lattices can be of different lengths, one having n_1 points and the other having n_2 points.

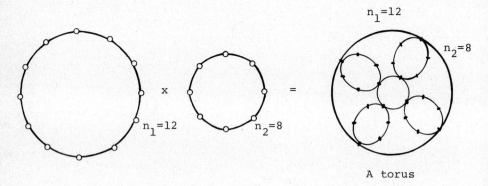

A torus

Fig. 9.4 Diagram of a torus like space.

Let f_{ij} be the probability that two genes separated by distance i along one direction and j along the other direction are identical by descent. Let

(9.28) $$F \equiv [f_{ij}]_{n_1 \times n_2} \equiv [F_0, F_1, \cdots, F_{n_2-1}]$$

where F_i are column vectors of dimension $n_1 \times 1$. Let

$$F_\boxtimes \equiv \begin{bmatrix} F_0 \\ F_1 \\ \cdot \\ \cdot \\ \cdot \\ F_{n_2-L} \end{bmatrix} \qquad \text{a column vector of } n_1 \times n_2 \text{ elements.}$$

Let M_1 and M_2 be the matrices corresponding to matrix M appearing in (9.11), and let

(9.29) $$M_\boxtimes^2 \equiv M_1 \boxtimes M_2$$

in which $\boxtimes$ means that if $A = [a_{ij}]_{nn}$ and $B = [b_{ij}]_{mm}$

(9.30) $$A \boxtimes B \equiv \begin{bmatrix} a_{11}B & a_{12}B & \cdots & a_{1n}B \\ a_{21}B & \cdots & & \\ \cdot & & & \\ \cdot & & & \\ \cdot & & & \\ a_{n1}B & \cdots & \cdots & a_{nn}B \end{bmatrix}_{nm \times nm}$$

(for the tensor product of vector spaces, see for example, Jacobson, (1953)).

Analogous to (9.7), we have

(9.31) $$F_{t+1}\boxtimes = (1 - u)^2 M_\boxtimes^2 [F_t\boxtimes + g_t\boxtimes]$$

where g_t given in (9.6), (Maruyama, 1970a).

A key theorem in treating these tensor products, as they are called, is that if $\{e_k, \lambda_k\}$ are the characteristic pairs of A, and $\{f_k, \gamma_k\}$ are of B, then the characteristic pairs of $A \boxtimes B$ are given by $\{f_k \boxtimes e_\ell, \lambda_k \lambda_\ell\}$. That is if

$$A e_k = \lambda_k e_k,$$
$$B f_k = \gamma_k f_k$$

then

$$(9.32) \qquad (A \boxtimes B)f_k \boxtimes e_\ell = \lambda_\ell \gamma_k f_k \boxtimes e_\ell \ .$$

Knowing property (9.32), we can analyse equation (9.31) in the same way as for the one dimensional problem.

If the migration rates in the two directions are respectively m_1 and m_2, then the equilibrium formula is

$$(9.33) \quad f_{ij} = \frac{(1-f_{00})(1-u)^2}{2Nn_1n_2} \sum_{k=0}^{\left[\frac{n_1+1}{2}\right]} \sum_{\ell=0}^{\left[\frac{n_2+1}{2}\right]} \frac{\Delta_k \Delta_\ell \xi_{k_\ell} \cos \frac{2i\pi k}{n_1} \cos \frac{2j\pi\ell}{n_2}}{\lambda_{k\ell}}$$

where

$$\xi_{k\ell} = \left\{ 1 - m_1 \left(1 - \cos \frac{2\pi k}{n_1} \right) \right\}^2 \left\{ 1 - m_2 \left(1 - \cos \frac{2\pi\ell}{n_2} \right) \right\}^2 ,$$

$$\lambda_{k\ell} = 1 - (1 - u)^2 \xi_{k\ell}$$

and

$$\Delta_0 = \Delta_{\frac{n}{2}} = 1 \quad \text{and} \quad \Delta_i = 2 \quad \text{otherwise} .$$

Also

$$f_{00} = \frac{(1-u)^2 S}{2N + (1-u)^2 S}$$

where

$$S = \frac{1}{n_1 n_2} \sum_{k=0}^{\left[\frac{n_1+1}{2}\right]} \sum_{\ell=0}^{\left[\frac{n_2+1}{2}\right]} \frac{\Delta_k \Delta_\ell \xi_{k\ell}}{\lambda_{k\ell}} \ .$$

9.4 Two-dimensional continuous space

Two-dimensional analogue of (9.25) is

$$(9.34) \quad f(x, y) = \frac{(1-u)^2(1-f_0)}{2DL_1L_2} \sum_{m=0}^{\infty} \sum_{n=0}^{\infty} \frac{\Delta_m \Delta_n R_{mn}}{1-(1-u)^2 R_{mn}} \cos \frac{2\pi mx}{L_1} \cos \frac{2\pi nx}{L_2}$$

where $\Delta_0 = 1$ and $\Delta_k = 2$ for $k \neq 0$, and

$$R_{mn} = \int_{-\infty}^{\infty} \int_{-\infty}^{\infty} \gamma(x, y) \cos \frac{2\pi mx}{L_1} \cos \frac{2\pi ny}{L_2} \, dxdy ,$$

(Maruyama, 1972a).

Also

$$f_0 = \frac{(1 - u)^2 S}{2DL_1L_2 + (1 - u)^2 S}$$

where

$$S = \sum_{m=0}^{\infty} \sum_{n=0}^{\infty} \frac{\Delta_m \Delta_n R_{mn}}{1 - (1 - u)^2 R_{mn}} .$$

In particular, if the migration obeys the normal distribution,

$$\gamma(x, y) = \frac{1}{2\pi\sigma^2} e^{-\frac{x^2+y^2}{2\sigma^2}}$$

then

$$R_{mn} = e^{-2\pi^2\sigma^2 \left\{ \frac{m^2}{L_1^2} + \frac{n^2}{L_2^2} \right\}} .$$

9.5 Higher order moments

The quantities discussed above are second order quantities and it can be considered as second moments. The higher moments can be also calculated in a similar way.

For simplicity, assume that the locus under consideration has K neutral alleles and that the mutation rate from one allele to any other allele is $u/(K-1)$. Then the expected equilibrium gene frequency is $1/K$. Considering the one dimensional circular case, let x_i be the gene frequency of a particular allele in colony i, $i = 1, 2, \ldots, n$. Now let $\xi_i = x_i - 1/K$. Thus ξ_i is the deviation of the frequency from the expectation. Let

$$P_t = \begin{pmatrix} \xi_{t, 1} \\ \xi_{t, 2} \\ \vdots \\ \xi_{t, n} \end{pmatrix} .$$

Where t indicates time, and let

$$S = (1 - m)I + \frac{m}{2}(R + R^{-1})$$

where m is the migration rate and R is the matrix defined on p.132. Then we have

$$P_{t+1} = (1 - u)S[P_t + \Delta_t] \tag{9.35}$$

where Δ_t is the vector consisting of the random changes due to sampling of gametes.

It is then easy to observe that

$$P_{t+1}\boxtimes^2 = (1 - u)^2 S\boxtimes^2[P_t + \Delta_t]\boxtimes^2 \tag{9.36}$$

where $\boxtimes^2$ indicates the tensor product of order 2 of the same vector or the same matrix. We also have

$$P_{t+1}\boxtimes^3 = (1 - u)^3 S\boxtimes^3[P_t + \Delta_t]\boxtimes^3 \tag{9.37}$$

and

$$P_{t+1}\boxtimes^4 = (1 - u)^4 S\boxtimes^4[P_t + \Delta_t]\boxtimes^4 \tag{9.38}$$

where $\boxtimes^3$ and $\boxtimes^4$ indicate the tensor products of order 3 and 4 respectively. Using the property that $(A + B)\boxtimes^2 = A\boxtimes^2 + 2A\boxtimes B + B\boxtimes^2$, equation (9.36) can be rewritten as

$$P_{t+1}\boxtimes^2 = (1 - u)^2 S\boxtimes^2[P_t\boxtimes^2 + \Delta_t\boxtimes^2]$$

because $P_t\boxtimes\Delta_t = \vec{0}$ = the zero vector. Vector $\Delta_t\boxtimes^2$ can be expressed as a linear combination of $P_t\boxtimes^2$ and $1\boxtimes P_t$. Similarly (9.37) and (9.38) can be reduced to a form involving $P_t\boxtimes 1$, $P_t\boxtimes^2$, $P_t\boxtimes^3$, etc.

9.6 Numerical analysis at equilibrium

It seems worthwhile to examine formulae (9.19), (9.22), (9.33) and (9.34) numerically. The dimension of the population space (one or two) makes a significant difference in the values of the probability of identity by descent, unless the mutation rate is the same order magnitudes as migration rate. The size of the space also has a large influence if $N_T u$ (the total population size times mutation rate) is small, which seems to be the case for many natural populations. The details of this situation can be revealed by numerical calculations using these formulae (see Table 9.1). Another quantity that can be studied from these formulae is the differences in local gene frequencies For instance if f_0, which is the probability of identity for two genes of zero distance, is large but $\bar{f}$ is small, we should expect a large amount of local differentiation, whereas if f_0 and $\bar{f}$ are both small and the ratio $\bar{f}/f_0$ is nearly unity, we may regard the population as approximately resembling a panmictic unit. However if both f_0 and $\bar{f}$ are large and nearly equal, we cannot predict anything from this fact alone. This is because if the mutation rate is low the population would be homallelic most of the time, and even if there is a large amount of local differentiation when it is polymorphic, this may contribute very little to the values of f_0 and $\bar{f}$. However, the ratio $(1-f_0)/(1-\bar{f})$ may be used to measure the effect of the geographical structure of a population in such a case. We should recall that $f_0 = \bar{f}$ for a panmictic population. (See Kimura and Maruyama, 1971.)

Using formulae (9.25) and (9.34) with the normal dispersion functions, these statistics were calculated for a number of cases. The numerical values obtained from the formulae reveal a biologically important fact: When $4N_T u < 1$, the ratio $(1-f_0)/(1-\bar{f})$ for a two-dimensional population is essentially determined by the value of $D(\sigma_x^2 + \sigma_y^2) = 2D\sigma^2$ alone where x and y are distances in the two directions, and is nearly independent of the habitat size, (it is assumed that $L_1 = L_2$ and $\sigma_x^2 = \sigma_y^2$). On the other hand, for a one-dimensional population, the ratio depends on both $2D\sigma^2$ and the habitat size. Furthermore, when $2D\sigma^2$ or $D(\sigma_x^2 + \sigma_y^2)$ is greater than 10, a two-dimensional population behaves very much like a panmictic unit, whereas with a one-dimensional population, even if $2D\sigma^2$ is large, the ratio can be made arbitrarily small by letting the habitat size be large. A few numerical examples illustrating this fact are presented in Table 9.1. In Cases 1 ∿ 3, 4 ∿ 6, and 7 ∿ 9 of Table 9.1, the ratios are nearly equal within each group despite the large differences in the total population number, in the population density, and in the habitat size. Cases 7 ∿ 9 show that a population with $D(\sigma_x^2 + \sigma_y^2) > 10$ exhibits very little local differentiation of the gene frequencies. It is remarkable that the value of the ratio $(1-f_0)/(1-\bar{f})$ is determined by the local property alone. Cases 10 ∿ 12 and 13 ∿ 15 are one-dimensional examples. They show clearly that the ratio depends on both $D\sigma^2$ and L. Therefore the local differentiation of the gene frequencies depends on the dimension of the habitat. Local differentiation may occur easily in one-dimensional cases, while two-dimensional populations tend to be more panmictic.

The value of f_0 is of biological interest because it is the probability of homozygosity. From the numerical examples, it was found that f_0 depends strongly on the habitat size when the population is two-dimensional and finite, particularly when $4N_T u$ is order of magnitude 1 or less. This is consistent with the fact just mentioned. Since a quasi-panmictic situation can be easily attained in two-dimensional finite cases, as the population becomes large and the genetic variability $(1-\bar{f})$ increases, the value of $(1-f_0)$ also increases accordingly. Therefore in a two-dimensional population of finite size, unless the migration is extremely restricted, there will not be a strong local differentiation of alleles. On the other hand, in one-dimensional cases, f_0 is nearly independent of the habitat size, unless the dispersion distance times density is greater than the length of habitat. Figure 9.5 illustrates typical patterns of the $f(r)$ in one- and two-dimensional spaces.

Table 9.1 Numerical examples of f_0, $\bar{f}$ and $(1-f_0)/(1-\bar{f})$, assuming that u (mutation rate) $= 10^{-7}$, and $\sigma_x^2 = \sigma_y^2$ in two-dimensional cases at equilibrium. Cases 1 ～ 9 are two-dimensional examples and the others are one-dimensional examples.

Case	Habitat size	Density	$D(\sigma_x^2 + \sigma_y^2)$ (two-dimensions) $D\sigma^2$ (one-dimension)	f_0	$\bar{f}$	$\dfrac{1 - f_0}{1 - \bar{f}}$
1	100 x 100	200	0.1	0.904	0.241	0.126
2	200 x 200	200	0.1	0.901	0.123	0.113
3	500 x 500	200	0.1	0.892	0.011	0.109
4	100 x 100	200	1.0	0.762	0.595	0.587
5	200 x 200	20	1.0	0.876	0.775	0.551
6	500 x 500	100	1.0	0.554	0.093	0.492
7	100 x 100	200	15.0	0.718	0.705	0.955
8	200 x 200	200	15.0	0.405	0.372	0.950
9	500 x 500	200	15.0	0.141	0.086	0.941
10	100	200	1.0	0.996	0.980	0.200
11	200	200	1.0	0.993	0.931	0.108
12	500	200	1.0	0.986	0.711	0.049
13	100	200	15.0	0.996	0.995	0.783
14	200	200	15.0	0.992	0.998	0.666
15	500	200	15.0	0.981	0.954	0.419

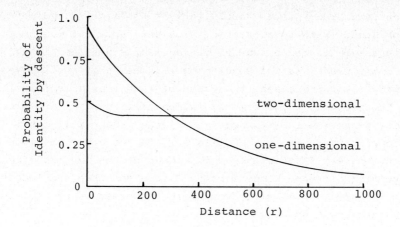

Fig. 9.5 Comparison of typical patterns in the relationship between the probability of identity by
descent and the geographical distance between two
genes in one- and two-dimensional habitats of
finite size, (4Nu is order of magnitude 1). In
the one-dimensional case, $L = 2000$, $2D\sigma^2 = 10$,
$D = 100$, $u = 10^{-7}$; in the two-dimensional case,
$L_1 = L_2 = 1000$, $D(\sigma_x^2 + \sigma_y^2) = 10$, $D = 1$, $u = 10^{-7}$.
Normal density and isotropicity are assumed for
the migration in both cases.

9.7 A differential equation and asymptotic formulae

Formulae (9.25) and (9.34) are the periodic solutions of the differential equation,

$$(9.39) \qquad \Delta f - 2uf + \left[\prod_{i=1}^{d} \delta(x_i)/2D \right] (1 - f) = 0$$

in which

$$\Delta = \sum_{i=1}^{d} \sigma_{x_i}^2 \frac{\partial^2}{\partial x_i^2} \quad ,$$

$\delta(x_i)$ is the delta function, and d is the number of dimensions of the
habitat. We can use this differential equation to investigate the
asymptotic behavior of the identity probability $f(r)$ discussed above.

In the one-dimensional case, we have simply

$$(9.40) \qquad \sigma_x^2 [d^2 f/dx^2] - 2uf + [\delta(x)/2D] = 0,$$

(Maruyama 1971).

Using the standard procedure of putting e^{mx} into the equation and determining m from the characteristic equation, we see that the solution is proportional to $\exp[-(x\sqrt{2u}/\sigma]$, which was first shown by Malécot (1948).

In higher dimension cases, if the migration is isotropic, we have

$$(9.41) \qquad \sum_{i=1}^{d} (\partial^2 f/\partial x_i^2) - (2u/\sigma^2)f = (1 - f/2D\sigma^2) \quad \text{times delta measure.}$$

By a transformation into polar coordinates, the above equation becomes

$$\frac{d^2 f(r)}{dr^2} + \frac{d-1}{r}\frac{df(r)}{dr} - \frac{2u}{\sigma^2} f(r) = \frac{1 - f(r)}{2D\sigma^2} \quad \text{times delta measure.}$$

If we let $y \equiv r \sqrt{(2u/\sigma^2)}$, this differential equation becomes

$$(9.42) \qquad \frac{d^2 f(y)}{dy^2} + \frac{d-1}{y}\frac{df(y)}{dy} - f(y) = \frac{1 - f(y)}{4Du} \quad \text{times delta measure.}$$

The left hand side of this equation does not contain any parameter except the dimension number, d, and this is a standard form. Mathematically, y is the standardized measurement of distance. Therefore $f(r)$ assumes an asymptotic form only if y is large or small, instead of the real distance, r. The solution of (9.42) can be expanded in terms of Hankel functions, and, in particular, the solution has a singularity at the origin, of the form

$$(9.43) \qquad \frac{1}{y^{d-2}} \quad \text{or} \quad \log\frac{1}{y}$$

depending on whether $d > 2$ or $d = 2$, and the solution is asymptotically proportional to

$$(9.44) \qquad e^{-y}y^{(1-d)/2} \quad \text{for } d \geq 1.$$

These asymptotic forms are valid for y, but not r alone. Taking logarithms, we have

$$-y + (\frac{1-d}{2})\log y .$$

In particular, if $d = 2$, this formula becomes

$$-y - \frac{1}{2} \log y .$$

Thus, if $\sqrt{2u/\sigma^2}$ is of order of magnitude 1 or greater, we should expect the identity probability to decrease exponentially with the distance, as pointed out by Imaizumi et al. (1971).

When $\sqrt{2u/\sigma^2}$ is small, it is incorrect to use an exponential formula like

(9.45) $f(r) = ae^{-br}$

to approximate the true $f(r)$, unless $d = 1$. If $d = 2$, this kind of exponential formula should be used only if $\sqrt{2u/\sigma^2}$ is large so that for all practical values of r, y is large and the asymptotic form (9.45) is valid. When $\sqrt{2u/\sigma^2}$ is small and $d = 2$, for small r, the slope of $f(r)$ near $r = 0$ is proportional to $\log(1/r)$, and there seems to exist no simple formula which is valid for the entire range of r. To confirm the validity of formulae (9.43) and (9.44), I have carried out extensive calculations of numerical values of $f(r)$ by computer. The results turned out to be in good agreement with the theoretical expectations. Namely, if $d = 1$ (one-dimensional) $f(r)$ can be well approximated in all cases by an exponential formula like (9.45). On the other hand, if $d = 2$ (two-dimensional) two essentially different situations exist. If the standardized distance $y = r\sqrt{(2u/\sigma^2)}$ is large an exponential formula fits well, but if $y = r\sqrt{(2u/\sigma^2)}$ is small, formula (9.43) is valid and a formula of the form

(9.46) $a + b \log (1/r)$

fits very well. The constants a and b of the formula were determined as follows: $f(1) = a$, $f(2) = a + b\log(1/2)$. Then the other values of $f(r)$ for the argument r greater than 2 were extrapolated. We should also be aware of the fact that no real population is infinite and therefore $\bar{f}$ itself fluctuates in time as well as the gene frequencies at any particular locality.

9.8   Random drift

So far, only stationary situations have been discussed. But the non-stationary problem of rate of decrease of heterozygosity due to random genetic drift is also of importance in population genetics. However, this problem is much harder to handle. As an application of the results obtained in the previous sections, we shall investigate the asymptotic rate of decrease (λ) of genetic variability.

It can be shown that the asymptotic rate is equal to

$$(9.47) \qquad \lambda = \frac{1}{2N_T} \lim_{u \to 0} \frac{1 - f_0}{1 - \bar{f}} = 2N_T \lim_{u \to 0} \frac{4N_T u (1 - f_0)}{4N_T u - (1 - f_0)} \, ,$$

(Maruyama, 1972).

Using a computer, I have calculated a large number of numerical values of λ from formula (9.47), and have found simple approximations:

$$(9.48) \qquad \lambda \approx \frac{\pi^2 \sigma^2 D}{L} \frac{1}{2N_T} \qquad \text{if } D\sigma^2 < \frac{L}{10} \quad \text{(one-dimensional)}$$

$$(9.49) \qquad \lambda \approx \frac{1}{2N_T} \qquad \text{if } D\sigma^2 > \frac{L}{10} \quad \text{(one-dimensional)}$$

$$(9.50) \qquad \lambda \approx \frac{\sigma^2}{2L^2} = \frac{D\sigma^2}{2N_T} \qquad \text{if } D\sigma^2 < 1 \quad \text{(two-dimensional)}$$

$$(9.51) \qquad \lambda \approx \frac{1}{2N_T} \qquad \text{if } D\sigma^2 > 1 \quad \text{(two-dimensional)}$$

As in the case of the identity probability, we see that with a one-dimensional population, whether it behaves like a panmictic population or not depends on both $D\sigma^2$ (a local property) and L (a global property), and also, for a fixed $D\sigma^2$, as L becomes large the rate λ always deviates from that of a random mating population. On the other hand, the rate λ depends only on $D\sigma^2$ in the two-dimensional case and if $D\sigma^2 > 10$, the population behaves nearly as a panmictic population and the rate is independent of the habitat size.

More exact analysis for one-dimensional problem was given by Nagylaki (1974), (see also Maruyama, 1970b, formula (2.7)).

In addition to the rate of decay discussed above, the shape of the eigenfunction associated with it is of considerable importance, because the eigenfunction gives information on the pattern of differentiation of the local gene frequencies. Let

$$(9.52) \qquad h(r) \equiv \lim_{u \to 0} c(1 - f(r))$$

in which $f(r)$ is given by (9.25) or (9.34), c is a normalization constant and u is mutation rate. For a one-demensional population, $f(x)$ is given by (9.25) and it is known that $h(x) \propto \sin \pi x/L$, if $D\sigma^2 \ll L$, (Maruyama, 1971). If the rate of decay is given by (9.48) or (9.50) we should expect strong local differentiation, and if it is given by

(9.49) or (9.51), we should expect the contrary. The following tables show some examples.

Table 9.2 Eigenfunction (one-dimensional),
the values of the function h(x) of (9.52).

$\frac{x}{L}$	$h(\frac{x}{L})$	
	Example 1	Example 2
0	0.0175	0.0180
0.05	0.0333	0.0329
0.10	0.0540	0.0535
0.15	0.0724	0.0723
0.20	0.0884	0.0886
0.25	0.1022	0.1023
0.30	0.1126	0.1136
0.35	0.1217	0.1223
0.40	0.1295	0.1295
0.45	0.1330	0.1333
0.50	0.1352	0.1336

The values of the parameters in example 1 are $L = 100$, $D = 1$, $\sigma^2 = 1$; the parameters in example 2 are $L = 1$, $D = 10$, $\sigma^2 = 0.002$. Function h(x) gives the asymptotic form of relative heterozygosity in the habitat.

Table 9.3 Eigenfunction (two-dimensional), the
values of the function h(x, y) of (9.52).

		$h(\frac{x}{L} , \frac{x}{L})$			
$\frac{x}{L}$	$D\sigma^2 =$ 100	10	1	0.1	0.01
0	0.0906	0.0866	0.0549	0.0111	0.0012
0.05	0.0908	0.0896	0.0835	0.0767	0.0763
0.10	0.0909	0.0905	0.0894	0.0886	0.0892
0.15	0.0909	0.0910	0.0926	0.0950	0.0961
0.20	0.0910	0.0914	0.0946	0.0991	0.1004
0.25	0.0910	0.0916	0.0960	0.1019	0.1032
0.30	0.0910	0.0917	0.0970	0.1038	0.1051
0.35	0.0910	0.0918	0.0976	0.1051	0.1063
0.40	0.0910	0.0919	0.0980	0.1059	0.1071
0.45	0.0910	0.0919	0.0982	0.1063	0.1075
0.50	0.0910	0.0919	0.0982	0.1064	0.1076

The values of the parameters are $L_1 = L_2 = 100$, $D = 10$.
With $D\sigma^2 \geq 10$, h(x, y) assumes nearly the same value
for every x and y which implies the genetic panmixture
of the population, while with $D\sigma^2 < 1$, h(0, 0) << h(x, y),
x, y $\neq$ 0 and therefore there will be the local differentiation of alleles.

The asymptotic rates of decay of genetic variability in structured populations are graphically compared with corresponding theoretical expectations. In Fig. 9.6, one dimensional circular populations
are treated. The two lines indicate the theoretical expectations,
$\lambda = 1/2N_T$ for larger dispersion and $\lambda = \sigma^2\pi^2/L^2$ for more restricted
dispersion. In Fig. 9.7, a two dimensional situation is shown.

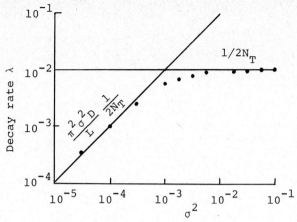

Fig. 9.6 One dimensional. The vertical axis stands
for the decay rate λ of (9.47). The horizontal
axis stands for the variance of migration distance (σ^2). L is the length of the habitat and
D is the population density. The two lines
represent the values of the two approximation
formulae (9.48) and (9.49). The dots indicate
the true value obtained by repeated iteration
of formula (9.24).

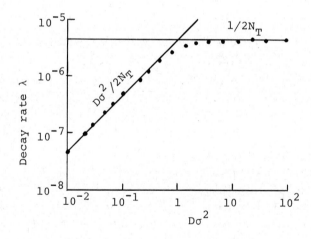

Fig. 9.7 Two dimensional. Two lines indicate the
approximations given by (9.50) and (9.51), while
the dots indicate the true values obtained from
(9.24). The values of the parameters are
$L_1 = L_2 = 100$ (habitat size), $D = 10$, and $N_T = DL^2$
(the total population number) $= 10^5$. Note that
the two approximation lines intersect at $D\sigma^2 = 1$.

9.9   Time to fixation

The asymptotic rate discussed above can be used to determine the time, T, (number of generations) required for a mutant gene to be established (fixed) by random drift, in the whole population, (Maruyama 1971a). The average time, $\bar{T}$, is approximately equal to $2/\lambda$, and the second moment of T is approximately equal to $6/\lambda^2$. It can be further shown that in the course of fixation the population stays for an equal number of generations at each intermediate gene frequency, e. g., in the case of two-dimensions, the population stays $2/(D\sigma^2)$ generations at each gene frequency. It is also possible to show that the probability density f(T) of time T is approximately equal to

$$(9.53) \qquad\qquad f(T) = \lambda^2 T e^{-\lambda T}$$

where λ is the asymptotic rate discussed above, (Maruyama, 1971a).

Table 9.4 and Fig. 9.8 are results of simulations which confirm the above assertions.

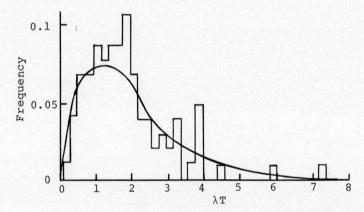

Fig. 9.8 Histogram of distribution of the fixation
time (T) from a simulation compared with the
theoretical expectation curve (9.53).   The
simulation consists of 100 fixation cases
divided into classes of interval $1/5\lambda$.   The
model is a circular population with n = 30,
2N = 2, and m = 0.2 (case 1 of Table 9.4).

Table 9.4 First and second moment of fixation time obtained from simulations are compared with theoretical expectations. μ_1 = the first moment (mean), and μ_2 = the second moment (not the variance).

Case	Moment	Theoretical Expectation	Simulation	Model	Number of Simulations used to compute the expectations
1	μ_1	900	869	Circular habitat with 30 colonies, 2N = 2, m = 0.2	100
	μ_2	1.2×10^6	1.0×10^6		
2	μ_1	5000	4122	Circular habitat with 50 colonies, 2N = 2, m = 0.1	85
	μ_2	3.8×10^7	2.2×10^7		
3	μ_1	5000	4650	Circular habitat with 50 colonies, 2N = 4, m = 0.1	95
	μ_2	3.8×10^7	2.9×10^7		
4	μ_1	800	856	Circular habitat with 50 colonies, 2N = 80, m = 0.1	24
	μ_2	9.6×10^5	9.7×10^5		
5	μ_1	2000	2072	Two-dimensional torus-like habitat with 10 x 10 colonies, 2N = 2, m = 0.1	100
	μ_2	6.0×10^6	5.7×10^6		
6	μ_1	534	531	Two-dimensional torus-like habitat with 10 x 10 colonies, 2N = 2, m = 0.7	47
	μ_2	4.3×10^5	3.4×10^5		

CHAPTER 10

GEOGRAPHICALLY INVARIANT PROPERTIES

In chapter 4 we dealt with the integration of various quantities along sample paths. There we assumed random mating. Here we will show that some of those integrations are independent of the population structure, even of structures much more general than those discussed in chapter 9. The invariant quantities are the fixation probability of a mutant gene with additive effects, the total number of heterozygotes counted during those generations when there is a given gene frequency for the whole population, and the sum of heterozygosity during the entire process. The higher moments of these quantities are also invariant.

Since the mathematics dealing with structured populations is not well established, I will present some of the invariant properties in a more elementary way. In section 10.1, the invariance of the sum of heterozygosity is shown for a discrete time model, and in section 10.2 a proof is given for the Moran model of continuous time. And in later sections, we will deal with diffusion equations.

10.1 Discrete time model

Model. Generation time is discrete. At the end of each generation the population is devided into nesting colonies where each colony produces exactly the same number of offspring as parents by random mating within the colony. Therefore the total population size (N) is constant over time. However, the number of colonies and the number of individuals in a colony may vary arbitrary over time.

Analysis. Let $N_i(t)$ be the number of individuals in colony i in generation t, and let $m_{ij}(t)$ be the probability that one born in colony i in generation t will participate in reproduction in colony j in the next generation, t+1. We assume that

(10.1)
$$\sum_j m_{ij}(t) = 1 \qquad \text{for all } i .$$

This assumes that the migration does not change the genetic composition in the entire population, though it will often change the local composition. Let $f_{ij}(t)$ be the probability that two randomly chosen homologous genes, one each from colony i and j, are the same. Define

(10.2)
$$F(t) \equiv \frac{1}{N^2} \sum_{ij} f_{ij}(t) N_i(t) N_j(t)$$

and

(10.3)
$$F_0(t) \equiv \frac{1}{N} \sum_i f_{ii}(t) N_i(t) .$$

Then $F(t)$ is the probability that two homologous genes randomly chosen from the entire population are the same allele in generation t, and $F_0(t)$ is the frequency of homozygotes. We assume no mutation during the time considered.

The fraction of colony i which comes from colony k is

$$\frac{m_{ki}(t) N_k(t)}{N_i(t+1)}$$

and

(10.4)
$$\frac{m_{ki}(t) N_k(t)}{N_i(t+1)} \frac{m_{\ell j}(t) N_\ell(t)}{N_j(t+1)} f_{k\ell}(t)$$

is the probability that two homologous genes, one each from colonies i and j, are the same allele and they come from colonies k and ℓ, provided $k \neq \ell$. With $k = \ell$, two homologous genes coming from a single colony are an identical gene in the previous generation with probability $1/(2N_k(t))$, and they are two different genes with probability $\{1 - 1/(2N_k(t))\}$. (Since we assume diploidy, there are $2N_k(t)$ genes rather than $N_k(t)$ genes which would be the case if haploid.) Thus the probability that two homologous genes in colonies i and j are the same allele and both come from colony k is

(10.5)
$$\frac{m_{ki}(t) N_k(t)}{N_i(t+1)} \frac{m_{kj}(t) N_k(t)}{N_j(t+1)} \left\{ \left(1 - \frac{1}{2N_k(t)}\right) f_{kk}(t) + \frac{1}{2N_k(t)} \right\} .$$

Summing over all possible combinations of k and ℓ in (10.4) and (10.5), we have

(10.6)
$$f_{ij}(t) = \sum_{k,\ell} \frac{m_{ki}(t)N_k(t)}{N_i(t+1)} \frac{m_{\ell j}(t)N_\ell(t)}{N_j(t+1)} f_{k\ell}(t)$$

$$+ \sum_k \frac{m_{ki}(t)N_k(t)}{N_i(t+1)} \frac{m_{kj}(t)N_k(t)}{N_j(t+1)} \frac{1 - f_{kk}(t)}{2N_k(t)} .$$

Multiplying the both sides of (10.6) by $N_i(t+1)N_j(t+1)/N^2$ and summing over i and j we have

(10.7)
$$F(t) = \frac{1}{N^2} \sum_{k,\ell} N_k(t) f_{k\ell}(t) \sum_{i,j} m_{ki} m_{\ell j}$$

$$+ \frac{1}{2N^2} \sum_k \{1 - f_{kk}(t)\} N_k(t) \sum_{i,j} m_{ki} m_{\ell j} .$$

Note that from (10.1)

$$\sum_{ij} m_{ki} m_{\ell j} = \sum_i m_{ki} \sum_j m_{\ell j} = 1 .$$

Therefore we have

(10.8)
$$F(t+1) = F(t) + \frac{1}{2N}\{1 - F_0(t)\} .$$

Thus

(10.9)
$$1 - \frac{1 - F(t+1)}{1 - F(t)} = \frac{1}{2N} \frac{1 - F_0(t)}{1 - F(t)} .$$

This formula was first given by Robertson (1964) for the general model in which the number of offspring is assumed to be exactly two for every individual.

Equation (10.9) can be rewritten as

$$2N\{F(t+1) - F(t)\} = 1 - F_0(t) .$$

The right side of the above equation is the frequency of heterozygotes, and thus the total sum (H) of the heterozygosity over time is given by

(10.10)
$$H = \sum_{t=0}^{\infty} \{1 - F_0(t)\} = 2N\{F(\infty) - F(0)\}$$

and

(10.11)
$$H = 2N\{1 - F(0)\} \qquad \text{if } F(\infty) = 1 .$$

Therefore the average of the total sum of heterozygosity which appears along the sample paths is independent of the population structure. In particular, if there is initially one mutant gene in the entire population, $F(0) = 1 - (1/N)$, and if complete splitting of the whole population does not occur, $F(\infty) = 1$,

(10.12) $$H = 2N\{1 - 1 + \frac{1}{N}\} = 2 .$$

Hence a singly present mutant gene without subsequent mutations produces on the average 2N heterozygote individuals before the mutant becomes extinct from the population or fixed in the population. And this property is independent of the population structure. We will show later that not only the average of the sum of the heterozygosity but the distribution of the sum among sample paths is also invariant with respect to the population structure, (Maruyama 1971b).

Heterozygosity and genetic variation. The quantity, $\Sigma x_i (1 - x_i)$ is a measure of genetic diversity or variation in a population, since it gives the probability of non-identity of two randomly chosen alleles. We shall call this quantity the "local genetic variability". In a diploid population this is also a measure of heterozygosity. We shall also refer to this quantity as heterozygosity in haploid populations, in analogy with the diploid case, even though there are no heterozygotes in a haploid population.

10.2 Continuous time model

This section provides a rigorous proof for some of invariant properties of a structured population (Maruyama 1974).

Model. A population consists of colonies between which migration is permitted. The organism is haploid and we consider a locus at which two alleles, say A and a, are segregating. When an individual dies, it is immediately replaced by an individual born in the same colony. Each preexisting individual in the colony has a chance of being chosen to give birth, which is proportional to that individual's fitness. The fitnesses of A and of a are 1+s and 1, respectively. The biological meaning of this assumption is that the local population density is tightly controlled by its space or food supply. It is assumed that each individual has the same negative exponential lifetime distribution. Between the death-birth events, migration between colonies or reformation of the colony structure by fusion and splitting may occur. There is no restriction on these processes. The gene frequencies of the whole population are not changed by this, but there will be local changes. This is a geographically structured version of Moran's model (1958) without mutation.

Analysis. Consider the particular moment at which a death-birth occurs. Let j be the number of A genes in the entire population and let j_ℓ be the number of A genes in colony ℓ. The probability that the death-birth occurs in colony ℓ whose size is N_ℓ, is N_ℓ/N. Given this, the probability that one a in colony ℓ dies is $(N_\ell - j_\ell)/N_\ell$ and the probability of birth of one A is $(1 + s)j_\ell/(N_\ell + sj_\ell)$. Therefore the conditional probability, $p_{j_\ell, j_\ell+1}$, that one a in the ℓth colony dies and is replaced by one A, is

$$(10.13) \qquad p_{j_\ell, j_\ell+1} = \frac{(1 + s)j_\ell(N_\ell - j_\ell)}{(N_\ell + sj_\ell)N_\ell} .$$

Similarly the conditional probability, $p_{j_\ell, j_\ell-1}$, that one A dies and replaced by one a, is

$$(10.14) \qquad p_{j_\ell, j_\ell-1} = \frac{j_\ell(N_\ell - j_\ell)}{(N_\ell + sj_\ell)N_\ell}$$

and the probability, p_{j_ℓ, j_ℓ}, that there is no change in the number of A individuals is

$$(10.15) \qquad p_{j_\ell, j_\ell} = 1 - p_{j_\ell, j_\ell+1} - p_{j_\ell, j_\ell-1} .$$

In the above, $\Sigma_\ell j_\ell = j$ and $\Sigma_\ell N_\ell = N$. Now we ask: what is the probability, $q_{j, j+1}$, that when a death-birth occurs, the number of A changes from j to j+1, given that the number is changed? It is

$$(10.16) \qquad q_{j, j+1} = \frac{\sum_\ell \frac{N_\ell}{N} p_{j_\ell, j_\ell+1}}{\sum_\ell \frac{N_\ell}{N} p_{j_\ell, j_\ell+1} + \sum_\ell \frac{N_\ell}{N} p_{j_\ell, j_\ell-1}}$$

$$= \frac{\sum_\ell \frac{j_\ell(N_\ell - j_\ell)(1 + s)}{N_\ell + sj_\ell}}{\sum_\ell \frac{j_\ell(N_\ell - j_\ell)(1 + s)}{N_\ell + sj_\ell} + \sum_\ell \frac{j_\ell(N_\ell - j_\ell)}{(N_\ell + sj_\ell)}}$$

$$= \frac{1 + s}{1 + s + 1} = \frac{1 + s}{2 + s} .$$

The corresponding probability, $q_{j,j-1}$, that the number of A changes from j to j-1, given that the number changes, can be similarly obtained:

(10.17)
$$q_{j,j-1} = \frac{1}{2+s} .$$

Note that the conditional probabilities $q_{j,j+1}$ and $q_{j,j-1}$ are independent of j and they are also independent of the geographical structure of the population. More generally, this holds even if the population structure changes. It holds even when the structure depends on the gene frequency, time, age of individual, etc.

Consider only those events in which the number of A is changed. Then the transition probabilities of this restricted process are given by $q_{j,j+1}$ and $q_{j,j-1}$ of (10.16) and (10.17). This means that the restricted Markov chain is a random walk on a set of integers [0, 1, 2, ..., N], whose transition probability of moving from i(i > 0, i < N) to i+1 is (1+s)/(2+s), and the probability of moving from i to i-1 is 1/(2+s). This procedure was first used, in population genetics, by Moran (1960), for a panmictic case. As a particular case, if s = 0, the Markov chain becomes a simple random walk on the interval.

Let r_{ij} be the total number of visits of the restricted process to state j before reaching the absorbing states 0 or N, provided that the starting state is i. We then have the well known solution,

$$R = [r_{ij}]_{(N-1) \times (N-1)}$$

which can be formally written as

(10.18)
$$R = [I - P]^{-1}$$

where I is the identity matrix, and

$$P = \begin{bmatrix} 0 & \lambda & & & & \mathbf{0} \\ \mu & 0 & \lambda & & & \\ & \mu & 0 & \cdot & & \\ & & \cdot & \cdot & \cdot & \\ \mathbf{0} & & & \cdot & \cdot & \cdot \\ & & & & \cdot & \end{bmatrix}_{(N-1) \times (N-1)}$$

where $\mu = (1+s)/(2+s)$ and $\lambda = 1/(2+s)$. The explicit formula is

$$(10.19) \qquad r_{ij} = \frac{2}{N} \sum_{k=1}^{N-1} \frac{(1 + s)^{(i-j)/2} \sin \frac{k\pi i}{N} \sin \frac{k\pi j}{N}}{1 - \frac{2(1 + s)^{1/2}}{2 + s} \cos \frac{k\pi}{N}} .$$

If s = 0

$$r_{ij} = 2i(1 - \frac{j}{N}), \qquad i \leqslant j ,$$

$$(10.20)$$

$$= 2j(1 - \frac{i}{N}), \qquad i > j .$$

The case of particular interest is that of i = 1:

$$(10.21) \qquad r_{ij} = 2(1 - \frac{j}{N}) .$$

When s ≠ 0, if we let N become large while keeping Ns = S constant,

$$r_{ij} \approx \frac{2N\{1 - e^{-S(1-j/N)}\}\{1 - e^{-Si/N}\}}{S(1 - e^{-S})} , \qquad i \leqslant j ,$$

$$(10.22)$$

$$\approx \frac{2N\{1 - e^{-S(1-j/N)}\}\{1 - e^{-Si/N}\}}{S(1 - e^{-S})}$$

$$- \frac{2N\{1 - e^{-S(i=j)/N}\}}{S} , \qquad i > j .$$

In particular, if i = 1, and if S >> 1 but s << 1,

$$r_{1j} \approx 2\{1 - e^{-S(1-j/N)}\} ,$$

$$(10.23)$$

$$\approx 2 \qquad \text{for} \quad S(1 - \frac{j}{N}) >> 1 .$$

Formulae (10.21) and (10.23) are probably the most important cases.

We have proved that the restricted process is independent of the population structure. This implies that the number of visits to state j is invariant with respect to population structure, and it is given by (10.18)-(10.23), provided that initially there are i A genes. This also proves that the ultimate fixation probability of an allele is not altered by the geographical structure of the population. The probability that A is ultimately established (fixed) in the whole population of size N is

$$\frac{1 - \left(\frac{1}{1 + s}\right)^i}{1 - \left(\frac{1}{1 + s}\right)^N} \approx \frac{1 - e^{-si}}{1 - e^{-sN}}$$

where i is the initial number of A. Moran (1960) derived this formula for a population without geographical structure (see also Kimura, 1957).

   There are other invariant properties. The following simple argument helps in deriving these properties. Let r_t be the probability that if the process is in a given state, the number of A genes changes during the t-th death-birth event (counted from beginning of this particular visit), provided that it did not change during the previous t-1 events. Then it is obvious that

(10.24) $r_1 + (1 - r_1)r_2 + (1 - r_1)(1 - r_2)r_3 + \cdots +$

        $(1 - r_1)(1 - r_2) \cdots (1 - r_n)r_{n+1} + \cdots = 1 .$

provided $\sum_n r_n = \infty$. In the case of a panmictic population in which s = 0,

$$r_k = \frac{2j(N - j)}{N^2} \qquad \text{for all k,}$$

and in the general geographically structured case,

$$r_k = \sum_\ell \frac{N_\ell}{N} P_{j_\ell, j_\ell + 1} + \sum_\ell \frac{N_\ell}{N} P_{j_\ell, j_\ell - 1}$$

$$= \frac{1}{N} \sum_\ell \frac{j_\ell(N_\ell - j_\ell)(2 + s)N_\ell}{N_\ell(N_\ell + sj_\ell)}$$

$$= \frac{1}{N} \sum_\ell \frac{j_\ell(N_\ell - j_\ell)N_\ell}{N_\ell(N_\ell + sj_\ell)/(2 + s)}$$

$$= \frac{1}{N} \sum_\ell \frac{2j_\ell(N_\ell - j_\ell)N_\ell}{N_\ell^2} \left[\frac{N_\ell(2 + s)}{2(N_\ell + sj)}\right]$$

in which N_ℓ and j_ℓ are measured just prior to the k-th death-birth event. Therefore, r_k is the quantity $2x_i(1 - x_i)$, where x_i is the local gene frequency at the moment of the k-th death-birth event, averaged over all colonies and weighted by $N_\ell(2+s)/2(N_\ell+sj)$. In a diploid population $2x_i(1-x_i)$ is the local heterozygosity.

   Consider a situation when the number of A genes is j. There will occur migration and death-birth events. We continue looking at this process until the number of A genes changes to j-1 or j+1. We now ask: What is the mean total amount of "heterozygosity" during this sequence of death-birth events until the number of A genes changes from j? The total amount would be $r_1+r_2+ \cdots +r_k$ with probability $(1-r_1)(1-r_2)\cdots (1-r_{k-1})r_k$. Therefore, the mean total heterozygosity during this time period is

$$(10.25) \quad r_1r_1 + (1 - r_1)r_2(r_1 + r_2) + (1 - r_1)(1 - r_2)r_3(r_1 + r_2 + r_3)$$
$$+ \cdots = r_1 + (1 - r_1)r_2 + (1 - r_1)(1 - r_2)r_3 + \cdots = 1 .$$

The last equality in the above equation follows from (10.24). Thus, at each visit to a particular state of the restricted process, the population sojourns there just long enough that on the average the mean heterozygosity becomes unity.

Since the number of visits to a particular state of the restricted process is independent of the geographical structure, the mean local heterozygosity summed over those death-birth moments at which the whole population is in a specified gene frequency is also independent. Furthermore, since each visit produces a mean total heterozygosity of unity, the mean total sum of heterozygosity accumulated while the whole population has a particular frequency for the gene A is the mean number of visits to that frequency. The number of visits is given by (10.18) -(10.23). It is important to note that the quantity shown to be nearly independent of the geographical stuucture is the sum of the local heterozygosity while the population is in a specified gene frequency range and not the sojourn time, the distribution of gene frequency, or the fixation time. In fact, all of the latter quantities are influenced by the population structure. Since most natural populations are geographically structured, this invariant property may prove to be useful.

The sum of the heterozygosity over all gene frequencies, that is, the total sum of mean local heterozygosity during the time when A and a are segregating, is also independent. The total sum is equal to $\Sigma_j r_{ij}$. With i = 1,

$$(10.26) \qquad \sum_j r_{1j} = 1, \qquad \text{if } s = 0 ;$$

$$(10.27) \qquad \sum_j r_{1j} \approx 2, \qquad \text{if } Ns \gg 1 \text{ but } s \ll 1 ;$$

and

(10.28) $\sum\limits_{j} r_{1j} \approx \frac{1}{-S}$,   if $Ns \ll -1$ but $|s| \ll 1$;

where $S = Ns$.

The higher moments of the quantities discussed above are also independent of the geographical structure of the population. Let $h_i^{(n)}$ be the n-th moment of the sum of heterozygosity, provided that the initial number of A is i. This is exactly the n-th moment of the sum of heterozygosity if $s = 0$, and this is approximately the n-th moment if $s \neq 0$. Let $h^{(n)} \equiv (h_1^{(n)}, h_2^{(n)}, \cdots, h_{N-1}^{(n)})^{*}$, where $*$ indicates the transpose of a vector. Then

(10.29) $$h^{(n)} = n(P - I)^{-1}h^{(n-1)} \qquad \text{for } n > 1 .$$

Let $r_{ij}^{(n)}$ be the n-th moment of the sum of heterozygosity, provided that the process starts from state i. Let $R^{(n)} \equiv [r_{ij}^{(n)}]_{(N-1) \times (N-1)}$, and let $R_{\delta}^{(n)} \equiv [\delta_{ij} r_{ij}^{(n)}]$ where $\delta_{ii} = 1$ and $\delta_{ij} = 0$ if $i \neq j$. Then

(10.30) $$R^{(n)} = n(P - I)^{-1}R_{\delta}^{(n-1)} \qquad \text{for } n > 1 .$$

10.3 Markov process

In the two preceding sections we showed that some of the quantities summed along the sample paths are invariant under the population structure. Here, using the model of section 10.2, we will show that the stochastic process of gene frequency change in a structured population can be shown to be a Markov process and furthermore that the Markov process is a simple random walk process and is independent of the population structure.

The model dealt in this section is the same model of N haploid individuals used in section 10.2.

Derivation of the Markov process. Let $N_{t,k}$ be the size of colony k at time t and X_t be the number of A genes in the entire population at time t. Then the random variable $\{X_t\}$ is a stochastic process whose state space is $[0, 1, 2, \cdots, N]$. This process with time parameter t can be very complicated, but we will introduce a new time measure which is sample-path-dependent local time based on an additive functional and which makes the process into a Markov process, actually a random time change discussed in section 5.12.

First, note that, in any short time interval, Δt, two or more death-birth events occur with probability of order $(\Delta t)^2$. We may therefore assume that the random variable X_t changes by at most one at any moment. Now consider a particular moment when a birth-death occurs, and ask what is the probability, given $X_t = i$ immediately before this event, that the value of X_t changes from i to i+1 through this event? Let i_k be the number of A genes in colony k at this moment, i.e., $i = \sum\limits_{k} i_k$.

The probability that the death-birth occurs in colony k whose size is $N_{t,k}$, is $N_{t,k}/N$, and given this, the probability q_{i_k,i_k+1}, that one a in the colony dies and it is replaced by one A, is

$$q_{i_k,i_k+1} = \frac{i_k(N_{t,k} - i_k)(1 + s)}{N_{t,k}(N_{t,k} + si_k)} ,$$

the same as (10.13). The probability, q_{i_k,i_k-1}, that one A dies and replaced by one a, is

$$q_{i_k,i_k-1} = \frac{i_k(N_{t,k} - i_k)}{N_{t,k}(N_{t,k} + si_k)} ,$$

the same as (10.14). The probability, q_{i_k,i_k}, that no change in the number of A occurs, is

$$q_{i_k,i_k} = 1 - q_{i_k,i_k+1} - q_{i_k,i_k-1}.$$

Therefore the probability, H(t), that the value of X_t changes through the death-birth event is

$$(10.31) \qquad H(t) = \frac{1}{N} \sum_{k} N_{t,k}[q_{i_k,i_k+1} + q_{i_k,i_k-1}]$$

$$= \frac{1}{N} \sum_{k} N_{t,k} \left[\frac{2i_k(N_{t,k} - i_k)}{N_{t,k}^2} \right] \left\{ 1 + \frac{s(\frac{1}{2} - \frac{i}{N})}{1 + \frac{is}{N}} \right\}.$$

H(t) can be defined for all t. We assume that migration takes place instantaneously and it is independent of death-birth events. The value of H(t) changes discontinuously as migration or a death-birth event occurs, and otherwise H(t) remains constant. For example, assume that the population consists of two colonies, say 1 and 2, and that each of the two colonies has two A genes and two a genes and s = 0. Under this situation, H(t) = (2/8)(4x2x2x2/16 + 4x2x2/16) = 1/2. Now suppose that one A gene moves from colony 1 to colony 2, then H(t) = (2/8)(3x1x2/9 + 5x3x2/25) = 7/15. It is important that the effects of

migration and of the population structure are incorporated in H(t).
Note that the quantity in the square bracket is the probability that
two gametes from a colony, randomly chosen with replacement, are different types. We call this quantity the local genetic variation. Therefore, H(t) is equal to the average local genetic variation if the
alleles are selectively neutral is approximately equal to it if there
is weak selection.

We now ask, if X_t is changed, what is the probability that X_t is
changed from i to i+1? This conditional probability is equal to

$$\frac{1}{H(t)} \frac{1}{N} \sum_k N_{t,k} q_{i_k, i_k+1} \cdot$$

Through simple algebra, this becomes

$$\frac{1 + s}{2 + s}$$

as in the case of (10.16). Similarly, the conditional probability
that X_t is changed from i to i-1 is

$$\frac{1}{2 + s} \cdot$$

We should note here that these quantities are independent of the geographical structure of the population.

Consider a short time interval Δt. Then the probability of occurrence of one death-birth event is

$$N\Delta t + o(\Delta t) \cdot$$

Let R_t be the probability that X_t is unchanged for time interval t.
Then

$$R_{t+\Delta t} = R_t (1 - N\Delta t) + R_t N\Delta t \left(1 - \frac{1}{\Delta t} \int_t^{t+\Delta t} H(\xi) d\xi \right) + o(\Delta t) \cdot$$

The first term on the right side of the above equation is the probability that X_t remains unchanged until time t and no death-birth occurs
in (t, t+Δt). The second term is the probability that X_t remains until
t and one death-birth occurs in (t, t+Δt) but X_t is unchanged. From
this equation we have

$$\frac{1}{R_t} \frac{dR_t}{dt} = -NH(t) ,$$

and therefore

(10.32)
$$R_t = e^{-N \int_0^t H(\xi)d\xi} .$$

Consider the stochastic process as a collection of sample paths $\{\omega\}$, and let

(10.33)
$$t \equiv \int_0^{\tau(\omega,t)} H(\omega, \xi)d\xi$$

where ω indicates a particular sample path and $H(\omega, \xi)$ is the same quantity as $H(\xi)$ of (10.31) associated with sample path ω. We assume that no part of the population is completely separated from the other. Consider this $\tau(\omega, t) = \tau$ as a new time measure. The τ is a local time based on the functional $H(t)$ of the original process. Because the τ depends on the history of the sample path, it is often called a stochastic clock. This is the kind of random time substitution discussed in section 5.12. We shall use the τ as the time parameter and construct a Markov process. Let X_τ be the number of A genes in the entire population at time τ. The X_τ is different from the X_t and it is not to be confused with the X_t. Let R_τ be the probability that X_τ is unchanged for time interval τ. Then, from (10.32) together with (10.33), we have

(10.34)
$$R_\tau = e^{-N\tau} .$$

This process X_τ is characterized as follows: (i) The state space is $[0, 1, 2, \cdots, N]$; (ii) The sojourn time at each state at each visit is exponentially distributed with mean $1/N$, provided $0 < i < N$. This follows from (10.34); (iii) The conditional probability that X_τ is changed from i to $i+1$ is $(1+s)/(2+s)$, and the conditional probability that X_τ is changed from i to $i-1$ is $1/(2+s)$, provided $0 < i < N$. These conditional probabilities are independent of the state and of the geographical structure of the population.

Let $q_{\tau,i,j}$ be the probability, given that X_τ is i at time 0, that it is j at time τ. From the properties (i)$\sim$(iii) above, we have

$$q_{\tau,i,j} = \sum_k q_{\tau-\xi,i,k} q_{\xi,k,j} \qquad \text{for } 0 < \xi < \tau,$$

and X_τ is a Markov process. Then we have, for small $\Delta\tau$,

$$q_{\tau-\Delta\tau,i,j} = e^{-N\Delta\tau} q_{\tau,i,j} + (1 - e^{-N\Delta\tau})[\lambda q_{\tau,i-1,j} + \mu q_{\tau,i+1,j}]$$

$$+ o(\Delta\tau) , \qquad 0 < i, j < N$$

where $\lambda = (1+s)/(2+s)$ and $\mu = 1/(2+s)$. From this equation, we derive

(10.35) $\qquad \dfrac{dq_{\tau,i,j}}{d\tau} = N[-q_{\tau,i,j} + \lambda q_{\tau,i-1,j} + \mu q_{\tau,i+1,j}], \quad 0 < i, j < N$

the KBE. Using the additive functional (10.33), this equation can be also derived from Dynkin's formula (1.70). We can derive the Kolmogorov forward equation similarly:

(10.36) $\qquad \dfrac{dq_{\tau,i,j}}{d\tau} = N[-q_{\tau,i,j} + \lambda q_{\tau,i,j-1} + \mu q_{\tau,i,j+1}]$.

These differential equations characterize the Markov process, and all the information concerning the process can be obtained from these equations. We define the following matrices:

$$Q_\tau \equiv [q_{\tau,i,j}]_{(N-1)\times(N-1)} \qquad 0 < i, j < N;$$

$$\frac{dQ_\tau}{d\tau} \equiv \left[\frac{dq_{\tau,i,j}}{d\tau}\right]_{(N-1)\times(N-1)},$$

$$A \equiv \begin{bmatrix} -1 & \mu & & & & \\ \lambda & -1 & \mu & & & \text{\Large O} \\ & \lambda & \cdot & \cdot & & \\ & & \cdot & \cdot & \cdot & \\ \text{\Large O} & & & \cdot & \cdot & \cdot \\ & & & & & -1 \end{bmatrix}_{(N-1)\times(N-1)}$$

where $\lambda = (1+s)/(2+s)$ and $\mu = 1/(2+s)$. Then the system of differential equations (10.35) becomes

(10.37) $\qquad \dfrac{dQ_\tau}{d\tau} = NAQ_\tau$

and (10.36) becomes

(10.38) $\qquad \dfrac{dQ_\tau}{d\tau} = NA^*Q_\tau$

where A^* is the transpose of A.

We shall next derive a diffusion process as a limit for large N. The KBE (10.35) can be rewritten as

(10.39)
$$\frac{dq_{\tau,i,j}}{d\tau} = \frac{1}{2} N[q_{\tau,i-1,j} - 2q_{\tau,i,j} + q_{\tau,i+1,j}]$$

$$+ \frac{Ns}{2(2+s)} [q_{\tau,i-1,j} - q_{\tau,i+1,j}] .$$

Let T be a new time measure such that $NT = \tau$, and $Ns = S$ where we assumes S is a constant. Let $q_N(T, \frac{i}{N}, \frac{j}{N}) \equiv q_{\tau,i,j}$. Then (10.39) can be written as

$$\frac{dq_N(T, \frac{i}{N}, \frac{j}{N})}{dT} = \frac{1}{2} \frac{1}{(\frac{1}{N})^2} \left[q_N(T, \frac{i-1}{N}, \frac{j}{N}) - 2q_N(T, \frac{i}{N}, \frac{j}{N}) \right.$$

$$\left. + q_N(T, \frac{i+1}{N}, \frac{j}{N}) \right] + \frac{Ns}{(2+s)} \frac{1}{(\frac{2}{N})} \left[q_N(T, \frac{i-1}{N}, \frac{j}{N}) \right.$$

$$\left. - q_N(T, \frac{i+1}{N}, \frac{j}{N}) \right] .$$

Therefore as $N \to \infty$, this difference equation converges to

(10.40)
$$\frac{\partial q_\infty(T, X, Y)}{\partial T} = \frac{1}{2} \frac{\partial^2 q_\infty(T, X, Y)}{\partial X^2} + \frac{S}{2} \frac{\partial q_\infty(T, X, Y)}{\partial X}$$

(see section 2.2). The Markov process governed by (10.40) is a Brownian motion.

Solutions. Based on the stochastic clock, whose time-rate depends on the total amount of the average local genetic variation which has appeared in the sample path, we have derived a Markov process of the gene frequency change. It is governed by differential equation (10.35), (10.36), (10.37), (10.38), or (10.40).

Now dealing with (10.37) and (10.40), we shall investigate biologically interesting questions. These equations together with definitions (10.31) and (10.33) tell us that, if we measure the time by τ the stochastic process becomes a Markov process that is independent of the geographical structure of the population and furthermore, that it is a random walk (or approximately a Brownian motion) on the finite interval [1, 2, $\cdots$, N-1], (or (0, 1)), with exit boundaries at both ends. A standard procedure in dealing with equation (10.37) is to obtain a complete spectral analysis of matrix A. The right eigenvectors of A are the column vectors

$$(10.41) \qquad e_k = \sqrt{\frac{2}{N}} \left((\frac{\lambda}{\mu})^{1/2} \sin \frac{\pi k}{N}, \ (\frac{\lambda}{\mu})^{2/2} \sin \frac{2\pi k}{N}, \right.$$

$$\left. (\frac{\lambda}{\mu})^{3/2} \sin \frac{3\pi k}{N}, \ \cdots, \ (\frac{\lambda}{\mu})^{(N-1)/N} \sin \frac{(N-1)\pi k}{N} \right)^*,$$

$$k = 1, 2, \cdots, N-1 \ .$$

where a* indicates the transpose of any row vector a. It can be easily shown that

$$A e_k = \lambda_k e_k$$

where

$$(10.42) \qquad \lambda_k = -1 + 2\sqrt{\lambda \mu} \cos \frac{\pi k}{N} \ .$$

Now let

$$g_k = \sqrt{\frac{2}{N}} \left((\frac{\mu}{\lambda})^{1/2} \sin \frac{\pi k}{N}, \ (\frac{\mu}{\lambda})^{2/2} \sin \frac{2\pi k}{N}, \ (\frac{\mu}{\lambda})^{3/2} \sin \frac{3\pi k}{N} \right.$$

$$\left. \cdots, \ (\frac{\mu}{\lambda})^{N-1/N} \sin \frac{(N-1)\pi k}{N} \right), \qquad k = 1, 2, \cdots, N-1 \ .$$

These row vectors are the left eigenvectors of A, i.e.,

$$g_k A = \lambda_k g_k$$

where the eigenvalues are the same as (10.42). Note that $\{e_k\}$, or $\{g_k\}$, themselves are not orthogonal systems, but $\{e_k\}$ and $\{g_k\}$ are biorthogonal, i.e.,

$$(e_k, g_\ell) = \frac{2}{N} \sum_{i=1}^{N-1} (\frac{\lambda}{\mu})^i \sin \frac{i\pi k}{N} (\frac{\mu}{\lambda})^i \sin \frac{i\pi \ell}{N} = \delta_{k\ell} \ .$$

where $\delta_{kk} = 1$ and $\delta_{k\ell} = 0$ if $k \neq \ell$.

With these eigenvectors and eigenvalues, we can easily obtain the fundamental solution of the process of equation (10.37), that is the transition probability:

$$Q_\tau = \frac{2}{N} \sum_{k=1}^{N-1} e^{-\lambda_k \tau} e_k \boxtimes g_k$$

where $\boxtimes$ indicates the direct product of two vectors. This can be written as

$$q_{\tau,i,j} = \frac{2}{N} \sum_{k=1}^{N-1} \exp\left\{-\tau(1 - 2\sqrt{\lambda\mu} \cos \frac{\pi k}{N})\right\} (\frac{\lambda}{\mu})^{(i-j)/2} \sin \frac{i\pi k}{N} \sin \frac{j\pi k}{N} .$$

The fundamental solutions of (10.40) are

$$q(T, X, Y) = 2e^{-S(X-Y)/2} \sum_{k=1}^{\infty} \exp\left\{-\frac{(k^2\pi^2+S^2)T}{8}\right\} \sin k\pi X \sin k\pi Y .$$

The Markov process governed by q(t, X, Y) is independent of the geographical structure and it is a Brownian motion.

The ultimate fixation probability, u_i, of allele A in the whole population is given by the solution of equation

$$Au = 0$$

where $u = (u_1, u_2, \cdots, u_{N-1})^*$, in which * indicates the transpose:

$$(10.43) \qquad u_i = \frac{1 - (\frac{1}{1+s})^i}{1 - (\frac{1}{1+s})^N} \approx \frac{1 - e^{-si}}{1 - e^{-sN}}$$

where i is the initial number of A genes in the whole population. This is the same as the formula in section 10.2. If we replace si and sN of (10.43) by Sx and S respectively, the right side in (10.43) is the solution of the differential equation

$$\frac{1}{2} \frac{d^2f}{dX^2} + \frac{S}{2} \frac{df}{dX} = 0$$

with boundary conditions f(0) = 0 and f(1) = 1 (Kimura 1962). The independence of the fixation probability here was derived under the assumption of constant size of the total population and local random mating, but importantly the structure of the population was not assumed to be fixed. Even if population structure depends on the genetic constitution, as would occur if genetically similar individuals tend to live closer to one another, or the structure depends on the gene frequency, the fixation probability is unaltered as proven also in section 10.2. This independence of the fixation probability was suggested by Maruyama (1970c), based on an approximation. However it was not clear at that time whether the independence is exact or an approximation.

The first exit time of the process from [1, 2, $\cdots$, N-1] is the solution of equation

$$Af + \underline{1} = \underline{0}$$

where $\underline{1} = (1, 1, \cdots, 1)*$ and $\underline{0} = (0, 0, \cdots, 0)*$. The i-th entry, f_i, of f is the mean duration of time, measured by τ, until allele A is fixed in the population or A is lost from the population:

$$f_i = \frac{2}{N} \sum_{j=1}^{N-1} \sum_{k=1}^{N-1} \frac{\left(\frac{\lambda}{\mu}\right)^{(i-j)/2} \sin \frac{i\pi k}{N} \sin \frac{j\pi k}{N}}{(1 - 2\sqrt{\lambda\mu} \cos \frac{\pi k}{N})} \qquad \text{for } 0 < i < N,$$

(see 10.19). If we let $f(\frac{i}{N}) \equiv f_i$ and let N become large while S = Ns stays constant, f(x) becomes the solution of differential equation

$$\frac{1}{2} \frac{d^2 f}{dx^2} + \frac{S}{2} \frac{df}{dx} + N = 0 :$$

(10.44)
$$f(x) = \frac{2}{s} \left[\frac{1 - e^{-SX}}{1 - e^{-S}} - x \right].$$

The biological meaning of f_i is that it is the sum of the average local genetic variation that appears in the population while the two alleles are segregating, provided that initially there are i A genes.

The second moment, $f_i^{(2)}$, of the first exit time measured by τ is the solution of

$$Af^{(2)} + 2f = 0$$

where f is the mean obtained above and $f^{(2)} = (f_1^{(2)}, f_2^{(2)}, \cdots, f_{N-1}^{(2)})$. The higher moments are similarly obtained:

$$Af^{(n)} + nf^{(n-1)} = 0.$$

Conditional paths. In the above calculations, we included both kinds of sample paths in which A is established in the whole population and in which A is lost. We can make a distinction between the two kinds, and calculate the quantities for those paths in which A is fixed, (or in which A is lost), as we did in section 5.9. Let f_i' be the mean exit time in the path in which A is fixed. Then $g \equiv (u_1 f_1', u_2 f_2', \cdots, u_{N-1} f_{N-1}')*$, is the solution of equation

$$Ag + u = 0$$

where u is the vector of the fixation probabilities, u_i. Biologically, f_i' is the expectation of the sum of the average local genetic variation given that A is established in the population. The solution is

$$f_i' = \frac{2}{u_i N} \sum_{j=1}^{N-1} \sum_{k=1}^{N-1} \frac{(\frac{\lambda}{\mu})^{i-j/2} \, u_j \, \sin \frac{i\pi k}{N} \sin \frac{j\pi k}{N}}{(1 - 2\sqrt{\lambda\mu} \cos \frac{\pi k}{N})} \qquad \text{for } 0 < i < N$$

where u_i is the fixation probability given in (10.33). The higher moments are the solution of

$$A g^{(n)} + n g^{(n-1)} = 0$$

where $g^{(0)} = u$ the fixation probability vector. As N becomes large, while $S = Ns$ remains constant, $f_1(\frac{i}{N}) = f_i'$ and $u(\frac{i}{N}) = u_i$ satisfy the differential equation

$$\frac{1}{u(X)} \left\{ \frac{1}{2} \frac{d^2 f_1(X)}{dX^2} + \frac{S}{2} \frac{df_1(X)}{dX} \right\} + N = 0 .$$

Finally we shall obtain the sojourn time at a given state, or in a given range of gene frequency. For $0 < i, j < N$, let Φ_{ij} be the sojourn time, measured by τ, at state j while A and a are segregating in the population, provided the process starts from state i. Let Φ_{1ij} be the sojourn time among the paths in which the fixation of A occurs. Let $\Phi = [\Phi_{ij}]_{(N-1)\times(N-1)}$ and $\Phi_1 = [\Phi_{1ij} u_i]_{(N-1)\times(N-1)}$. Then these matrices are

$$\Phi = -A^{-1}$$

and

$$\Phi_1 = -A^{-1} U$$

where $U = [\delta_{ij} u_i]$ with $\delta_{ii} = 1$ and $\delta_{ij} = 0$ if $i \neq j$.

$$\Phi_{ij} = \frac{2}{N} \sum_{k=1}^{N-1} \frac{(\frac{\lambda}{\mu})^{i-j/2} \sin \frac{i\pi k}{N} \sin \frac{j\pi k}{N}}{(1 - 2\sqrt{\lambda\mu} \cos \frac{\pi k}{N})} \qquad 0 < i, j < N,$$

$$\Phi_{1ij} = \frac{2}{u_i N} \sum_{k=1}^{N-1} \frac{u_j (\frac{\lambda}{\mu})^{i-j/2} \sin \frac{i\pi k}{N} \sin \frac{j\pi k}{N}}{(1 - 2\sqrt{\lambda\mu} \cos \frac{\pi k}{N})} \qquad 0 < i, j < N.$$

As N becomes large,

$$\Phi_{ij} \approx \frac{(1 - e^{-2s(N-j)})(1 - e^{-2si})}{s(1 - e^{-2Ns})} \qquad \text{if } j > i,$$

(10.45)

$$\Phi_{ij} \approx \frac{1}{s} \left\{ \frac{(1 - e^{-2s(N-j)})(1 - e^{-2si})}{(1 - e^{-2Ns})} - (1 - e^{-2s(i-j)}) \right\}$$

$$\text{if } j < i.$$

If s = 0

$$\Phi_{ij} = \frac{2i(N - j)}{N} \qquad \text{if } j > i,$$

(10.46)

$$\Phi_{ij} = \frac{2j(N - i)}{N} \qquad \text{if } j < i.$$

Formula (10.46) is known for a panmictic population, (cf. Ewens 1963). If we let $\Phi(x, y) = \Phi(\frac{i}{N}, \frac{j}{N}) = \Phi_{ij}$, this satisfies the following differential equation

$$\frac{1}{2} \frac{d^2 \Phi(X, Y)}{dX^2} + \frac{S}{2} \frac{d^2 \Phi(X, Y)}{dX} + N\delta(X - Y) = 0 ,$$

where $\delta(\cdot)$ is the Dirac delta function. For large N, Φ_{1ij} can be obtained as the solution of

$$\frac{1}{2} \frac{d^2 \Phi_1(X, Y)}{dX^2} + \frac{S}{2} \frac{d\Phi_1(X, Y)}{dX} + Nu(X) \delta(X - Y) = 0 .$$

Approximation formulas (10.44), (10.45), and (10.46) were obtained in Maruyama (1972), where differential equation (10.40) and the method described in Dynkin (1965, vol.2, pp. 46-53) were used. Before concluding this section, I would like to emphasize that the sojourn time, measured by the number of generations, the time until fixation of an allele, the distribution of the gene frequencies and the rate of decay of genetic variability, depend upon the population structure.

10.4 Diffusion method

Here we assume that the organism is diploid. We consider a population subdivided into colonies and let $N_{t,i}$ be the size of colony i ($2N_{ti}$ genes). We assume that the total population size

$$N \equiv \sum_i N_{t,i}$$

is constant, that selection and random mating occur independently in each colony, and that no part of the population is completely separated from the others. We consider a locus at which two alleles A and a are segregating and denote by $x_{t,i}$ the frequency of A in colony i at time t. Let the fitness of AA, Aa and aa measured in Multhusian parameters be 2s, s and 0 respectively.

Now let

$$X(t) \equiv \frac{1}{N} \sum_i N_{t,i} x_{t,i} ,$$

be the gene frequency in the entire population at time t. We assume as before that neither mutation nor migration from outside of the population occurs during the time considered. The change of the X(t) is the stochastic process to be investigated by the diffusion method. The diffusion process of X(t) is governed by the two quantities: The variance $(V_{\delta x}(t))$ and the mean $(M_{\delta x}(t))$ of the change in X(t) in one generation. We shall now calculate $V_{\delta x}(t)$ and $M_{\delta x}(t)$. Note that

$$X(t)' = \frac{1}{N} \sum_i N_{t,i} x_{t,i} = \frac{1}{N} \sum_i N_{t,i} \{x_{t,i} + \Delta x_{t,i}\} ,$$

where the primes indicate the quantity after the random sampling of gametes but before migration and $\Delta x_{t,i}$ is the fluctuation in $x_{t,i}$ due to the sampling. By the assumption of independent sampling in each colony,

$$E\{\Delta x_{t,i} \Delta x_{t,j}\} = 0 \qquad \text{if } i \neq j$$

and

$$E\{\Delta x_{t,i}^2\} = \frac{x_{t,i}\{1 - x_{t,i}\}}{2N_{t,i}}$$

in which $E\{\cdot\}$ stands for the expectation over all sample paths. Therefore we have

(10.47) $$V_{\delta X}(t) \equiv E\{[X(t)' - X(t)]^2\} + \{E[X(t)' - X(t)]\}^2$$

$$= \frac{1}{2N^2} \sum_i x_{t,i}\{1 - x_{t,i}\}N_{t,i} + O(s^2) .$$

The assumption of independent selection in each colony implies

$$E\{\Delta x_{t,i}\} \equiv E\{x'_{t,i} - x_{t,i}\} = s x_{t,i}\{1 - x_{t,i}\} ,$$

where the prime indicates the quantity after the selection. Thus

$$M_{\delta X}(t) \equiv E\{X'(t) - X(t)\} = E\{\frac{1}{N} \sum_i \Delta x_{t,i} N_{t,i}\}$$

(10.48)
$$= \frac{s}{N} \sum_i x_{t,i}\{1 - x_{t,i}\}N_{t,i} \quad .$$

Now let

(10.49)
$$H(X(t)) = \frac{2}{N} \sum_i x_{t,i}\{1 - x_{t,i}\}N_{t,i} \quad .$$

Let $P(t, X, Y)$ be the probability density that the $X(t)$ moves from X to Y in time interval t. Since migration may change the local distribution of genes, and colony size, it may change $M_{\delta X}(t)$ and $V_{\delta X}(t)$. However, the global gene frequency $X(t)$ is not altered by migration. Thus the $P \equiv P(t, X, Y)$ satisfies the following KBE

(10.50)
$$\frac{\partial P}{\partial t} = \frac{V_{\delta X}(t)}{2} \frac{d^2P}{dx^2} + M_{\delta X}(t)\frac{dP}{dX} \quad .$$

This is a diffusion process with time dependent coefficients. Let

$$A(t) \equiv \frac{V_{\delta X}(t)}{2} \frac{d^2}{dx^2} + M_{\delta X}(t)\frac{d}{dX} \quad .$$

Then KBE (10.50) becomes

(10.51)
$$\frac{dP}{dt} = A(t)P \quad .$$

It is impossible to deal with KBE (10.50) in this form, for the coefficients $V_{\delta X}(t)$ and $M_{\delta X}(t)$ are expressed in terms of the local gene frequencies. Note that the two coefficients given in (10.48) and (10.49) have the factor

$$\sum x_{t,i}\{1 - x_{t,i}\}N_{t,i} \quad ,$$

hence it is possible to apply a random time change discussed in section 5.12.

Following (5.33), let

(10.52)
$$t = \int_0^{\tau_\omega(t)} H(X_\omega(\xi))d\xi$$

where H(X) is the quantity given by (10.49), and ω indicates the sample path. With this definition, the new time parameter τ is the sum of the quantity

(10.53)
$$\frac{2}{N} \sum_i x_{t,i} \{1 - x_{t,i}\} N_{t,i}$$

along each sample path. We will refer to this as the sum of the heterozygotes, because each term in (10.53) is the heterozygote of population i. Therefore the speed of τ can be different depending on the sample path. Mathematically, τ_ω is a nonnegative additive functional of the process and thus this can be used as a time parameter, see section 10.3. Using τ_ω, let $Q(\tau, X, Y)$ be the probability density that the gene frequency of A goes from X at $\tau_\omega = 0$ to Y at $\tau_\omega = \tau$. Then equation (10.50) becomes

(10.54)
$$\frac{\partial Q}{\partial \tau} = \frac{1}{8N} \frac{\partial^2 Q}{\partial X^2} + \frac{s}{2} \frac{\partial Q}{\partial X}$$

where $Q = Q(\tau, X, Y)$. (The procedure employied to reduce equation (10.50) to (10.54) is entirely different from the arc-sine transformation invented by Fisher 1930; see 1958, p. 96. His method is a transformation of the space variable to make the drift coefficient constant, while the method here is a time substitution which depends on the individual sample path.) Equation (10.54) has a two-fold consequence. Namely, (i) the process governed by $Q(\tau, X, Y)$ is a Brownian motion, and (ii) it is independent of the population structure, as shown in section 10.3.

The appropriate solution of (10.54) is

(10.55) $\quad Q(\tau, X, Y) = 2e^{-2S(X-Y)} \sum_{n=1}^{\infty} e^{-(n^2\pi^2+4S^2)\tau/8N} \sin n\pi X \sin n\pi Y$,

which is essentially the same as fundamental solution $q_\infty(T, x, y)$ given in section 10.3.

10.5 Computer simulation of gene frequency change

To show the validity of formula (10.55) several computer simulations have been performed. The results are presented in Fig. 10.1 and 2. One simulation was done with a panmictic population of size 20. The other was done with a population subdivided into 4 colonies, circularly arranged, each having 5 individuals, and adjacent colonies exchanging their members at a rate of 0.01. In both cases, the

initial gene frequency was set to 0.25. The initial distribution in
the subdivided population was $x_{0,1} = 1.0$ and $x_{0,2} = x_{0,3} = x_{0,4} = 0$.
To show the effect of the population structure, both $P(t, X, Y)$ and
$Q(\tau, X, Y)$ are presented. It is remarkable that $Q(\tau, X, Y)$ of these
two situations coincide in Fig. 10.1 despite the large difference of
$P(t, X, Y)$ in Fig. 10.2. Spreading of the gene frequencies among populations is greatly delayed in the subdivided case, (Fig. 10.2).

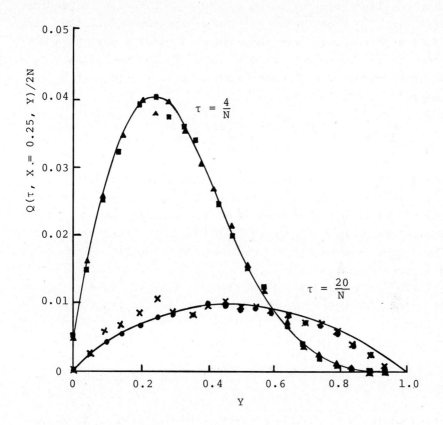

Fig. 10.1 The probability distributions of the gene frequencies among populations, $Q(\tau, X, Y)$. Curves indicate the theoretical expectations calculated from
formula (10.55) with $s = 0$ and points indicate the
simulation results (each point is the average of about
30,000 repetitions); ■ and ● represent the results
of the panmictic case, ▲ and ✗ represent the subdivided case.

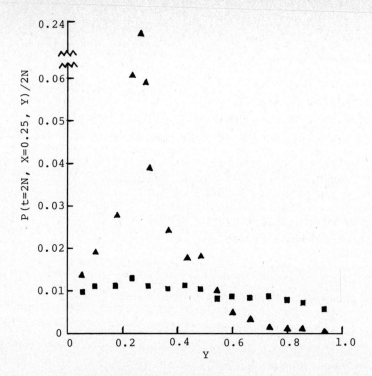

Fig. 10.2 The original process, P(t, X, Y). ■ represents the panmictic case and ▲ the subdivided case.
Note the difference of the spreading of the gene
frequencies in the two cases.

10.6 Invariance based on diffusion method

Let us now investigate various path integrals of the process
governed by $Q(\tau, X, Y)$. The ultimate fixation probability is not
altered by going from the process governed by $P(t, X, Y)$ to the process governed by $Q(\tau, X, Y)$, because in this transformation only the
time scale is changed, but the road map remains exactly the same.
Therefore the fixation probability $u(X)$ can be obtained by solving the
equation,

$$Au(X) = \frac{1}{8N} \frac{d^2 u(X)}{dX^2} + \frac{s}{2} \frac{du(X)}{dX} = 0$$

with boundary conditions $u(0) = 0$ and $u(1) = 1$. The appropriate solution is

$$u(X) = \frac{1 - e^{-4NsX}}{1 - e^{-4Ns}} ,$$

which is identical to (4.64) and essentially the same as the results shown in sections 10.2 and 10.3.

Next consider the time T(X) to fixation or extinction, according to (4.6) and (4.7), T(X) satisfies

(10.56) $\qquad AT(X) + 1 = \frac{1}{8N} \frac{d^2T(X)}{dx^2} + \frac{s}{2} \frac{dT(X)}{dX} + 1 = 0$

with the boundary conditions $T(0) = T(1) = 0$ (see (4.6) with $f(X) = 1$). The appropriate solution of (10.56) is

(10.57) $\qquad T(X) = \frac{2(1 - e^{-4NsX})}{s(1 - e^{-4Ns})} - \frac{2X}{s}$

(10.58) $\qquad\qquad = \frac{2(u(X) - X)}{s} \qquad$ if $s \neq 0$

where $u(X)$ is the fixation probability. If $s = 0$,

(10.59) $\qquad\qquad T(X) = 4NX(1 - X) ,$

and in particular if $X = 1/2N$,

(10.60) $\qquad\qquad T(\frac{1}{2N}) = 2(1 - \frac{1}{2N}) \approx 2 .$

For $s \neq 0$, if $4Ns \gg 1$ and $s \ll 1$,

(10.61) $\qquad\qquad T(\frac{1}{2N}) \approx 4$

and if $4Ns \ll -1$ and $|s| \ll 1$,

(10.62) $\qquad\qquad T(\frac{1}{2N}) \approx \frac{-1}{Ns} .$

Although mathematically the quantity given by these formulae is the average time a sample path stays in (0,1) before it goes to fixation or extinction measured in the units of τ, biologically they are the average of the sum of the heterozygotes which appear in the population (see the definition of the time parameter τ, (10.52)). Using the theory developed in chapter 4, we can also obtain higher moments of T(X).

The second moment is given by

$$T^{(2)}(X) = \frac{32NS(X)}{sS(X)} \left\{ \frac{1 + e^{-S}}{SS(1)} - \frac{2}{s^2} - \frac{1}{2S} + \frac{1}{s^2} - \frac{S(1)}{s^3} \right\}$$

$$- \frac{32N}{s} \left\{ \frac{X(1 + e^{-SX})}{SS(1)} - \frac{2S(X)}{s^2S(1)} - \frac{X^2}{2S} + \frac{X}{s^2} - \frac{S(X)}{s^3} \right\}$$

if $s \neq 0$, where $S = 4Ns$ and $S(X) = 1 - \exp(-4NsX)$. If $s = 0$,

$$(10.63) \qquad T^{(2)}(X) = \frac{16N^2(X - 2X^3 + X^4)}{3} \quad .$$

We shall next consider the sojourn time distribution of $Q(\tau, X, Y)$. Let $\phi^{(1)}(X, Y)dY$ be the average time (measured in τ) spent in $(Y, Y+dY)$. Then $\phi^{(1)}(X, Y)$ satisfies

$$(10.64) \qquad A\phi^{(1)}(X, Y) + \delta(X - Y)$$

$$= \frac{1}{8N} \frac{d^2\phi^{(1)}(X, Y)}{dX^2} + \frac{s}{2} \frac{d\phi^{(1)}(X, Y)}{dX} + \delta(X - Y) = 0$$

where $\delta(\cdot)$ is Dirac's delta function, (see (4.55) and (4.56)). The solution is

$$(10.65) \quad \phi^{(1)}(X, Y) = \frac{2(1 - e^{-4Ns(1-Y)})(1 - e^{-4Ns})}{s(1 - e^{-4Ns})} \qquad \text{for } X < Y,$$

$$= \frac{2(1 - e^{-4Ns(1-Y)})(1 - e^{-4Nsx})}{s(1 - e^{-4Ns})}$$

$$- \frac{2(1 - e^{-4Ns(X-Y)})}{s} \qquad \text{for } X > Y.$$

If $s = 0$, this reduces to

$$(10.66) \qquad \phi^{(1)}(X, Y) = 8N(1 - Y)X \qquad\qquad \text{for } X < Y,$$

$$= 8N(1 - Y)X - 8N(X - Y) \qquad \text{for } X > Y$$

and the particular case of $X = 1/2N$ is

$$\phi^{(1)}(\frac{1}{2N}, Y) = 4(1 - Y) \qquad \text{for } \frac{1}{2N} < Y < 1 \; .$$

If s in the above formula is small but positive and 4Ns >> 1, it becomes the following very simple distribution

(10.67)
$$\Phi^{(1)}(\frac{1}{2N}, Y) \approx 4 .$$

It is worth noting that, if 4Ns is large, the expected sum of heterozygotes is 4N and the expected conditional sum of heterozygosity for all given $Y = 1/2N, 2/2N, \cdots$, is 4/2N and this quantity is independent of Y, while if s = 0 the expectation is 2N and the conditional heterozygosity is given by the density 4(1-Y)/2N. If 4Ns << -1 and $|s| << 1$,

(10.68)
$$\Phi^{(1)}(\frac{1}{2N}, Y) \approx \frac{4}{e^{-S}} S(1 - Y) .$$

The second moment ($\Phi^{(2)}(X, Y)$) of the quantity given by $\Phi^{(1)}(X, Y)$ of (10.65) is also invariant under the geographical structure and is the solution of the differential equation

$$A\Phi^{(2)}(X, Y) + 2\delta(X - Y)\Phi^{(1)}(X, Y) = 0$$

where $\Phi^{(1)}(X, Y)$ is given in (10.65). Thus

(10.69)
$$\Phi^{(2)}(X,Y) = \frac{16NS(1-Y)e^{-BX}S(1-X)\Phi^{(1)}(Y,Y)}{BS(1)} \qquad \text{for } Y > X,$$

$$= \frac{16N\Phi^{(1)}(Y,Y)}{B}\left\{\frac{S(1-Y)e^{-BX}S(1-X)}{S(1)} - e^{BY}S(X)\right\}$$

$$\text{for } Y < X .$$

There are some correspondence between formulae in sections 10.2, 10.3 and 10.4. Formulae (10.22), (10.45) and (10.65) are equivalent. Formula (10.44) corresponds to (10.57).

These analyses make no distinction between the sample paths going to fixation or extinction of the allele in question. We can apply the method studied in section 5.2 and choose conditional paths destined to either fixation, extinction or a particular gene frequency. Let $T_1^{(1)}(X)$ and $T_1^{(2)}(X)$ be the average sum and the second moment of the time to fixation measured by τ, which is the sum of heterozygosity along those paths destined to fixation. Then these functions satisfy the equation

(10.70) $$AT_1^{(n)}(X) + nT_1^{(n-1)}(X) = 0$$

where $A = u^{-1}(X)Au(X)$. The special case of $s = 0$, we have

(10.71) $$T_1^{(1)}(X) = \frac{4N(1 - X^2)}{3}$$

and

(10.72) $$T_1^{(2)}(x) = \frac{112N^2}{45} + \frac{32N^2}{3} \frac{x^4}{10} - \frac{x^2}{3} .$$

Therefore, regardless the population structure, if a neutral mutant is fixed by random drift, it produces $4N^2/3$ heterozygotes on the average. We can also obtain the expected sum of heterozygotes for the extinction cases by solving equation

$$\frac{4N^2}{3} \frac{1}{2N} + (1 - \frac{1}{2N})x = 2N$$

which turn out to be $x = 4N/3$. Hence we can conclude that of the expected sum $(2N)$ of heterozygotes due to a single mutant, one-third $(1/2N \times 4N^2/3)$ will occur in populations in which it is eventually fixed, and two-thirds $((1-1/2N) \times 4N/3$, approximately) in those in which it is lost.

10.7 Computer simulation of heterozygote distribution and other invariant properties

Several computer simulations were carried out using the following three models of population structure: Model I consists of ten circularly arranged colonies of equal size, and geographically adjacent colonies exchange their members at the rate m; Model II consists of ten colonies of variable size (but the total size is fixed) and an individual moves randomly from one colony to any other colony with probability m per generation; Model III is a random mating population. The generations are discrete in all three models and a mutant gene is introduced into the population when and only when it becomes homallelic. Six examples of such comparisons are presented in the table and the figure below. In the table, the mean and the second moment of the sum of heterozygosity due to a single mutant gene are compared with the theoretical expectations under various conditions, and in order to show the structural difference, the mean fixation time of mutants is also given. In the figure, the distribution of the sum of heterozygosity under specified gene frequencies are compared. It is quite remarkable that despite large differences in the fixation times due to

Table 10.1 Simulation results ($T^{(1)}(X)$, $T^{(2)}(X)$, $T_1^{(1)}$ and $u(X)$) are compared with their theoretical expectations.

(In all the simulations given here, the initial frequency was set to $1/2N$ (= X). Theoretical expectations are given in parentheses. The fixation time is given for purpose of showing the effect of population structure.)

Case	N*	m**	s***	$T^{(1)}$	$T^{(2)}$	$T_1^{(1)}$	u†	T‡	Model§
1	100	0.01	0.05	3.88(3.81)	150.4(136.3)	37.3(36.0)	199/2000=0.0995(0.0952)	627.0	I
2	100	0.10	0.05	3.75(3.81)	145.6(136.3)	34.0(36.0)	170/1594=0.1067(0.0952)	396.1	II
3	100	—	0.05	3.76(3.81)	153.3(136.3)	37.9(36.0)	200/2122=0.0943(0.0952)	141.6	III
4	50	0.02	0	1.88(2.00)	120.0(133.3)	60.2(66.6)	62/7000=0.0089(0.01)	842.9	I
5	50	0.06	0	2.05(2.00)	140.3(133.3)	62.6(66.6)	71/6972=0.0102(0.01)	342.3	II
6	50	—	0	1.78(2.00)	108.9(133.3)	59.8(66.6)	200/20936=0.0096(0.01)	174.4	III

* N = population size. ** m = migration rate. *** s = selection coefficient.

† Fixation probability is given by the number of fixations occurred/ the number of runs.

‡ Fixation time. § See the text for the models.

186

the population structure, as shown in the table, the sum of heterozygotes and the distribution of heterozygosity are invariant over the structure.

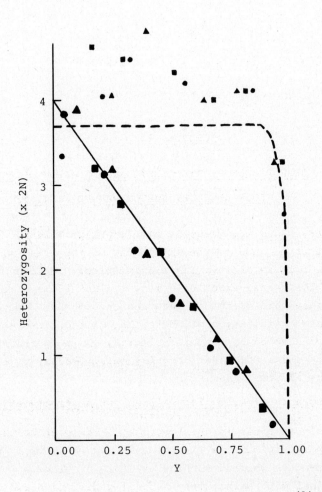

Fig. 10.3 The distribution heterozygosity $\Phi^{(1)}(X,Y)$, with X(initial frequency) = 1/2N and Y the gene frequency in the entire population). The dots indicate simulation results. The simulation data were taken from those presented in the table. • = case 1; ▲ = case 2; ■ = case 3; ● = case 4; ▲ = case 5; ■ = case 6. The broken curve indicates the theoretical expectation for simulations nos. 1-3, and the solid line indicates the expectation for simulations nos. 4-6.

CHAPTER 11

GENE FREQUENCY DISTRIBUTIONS AND RANDOM DRIFT IN

GEOGRAPHICALLY STRUCTURED POPULATIONS

In chapters 9 and 10, we dealt with problems that permit exact
solutions, either because of the simplicity of the problem, or of the
population, or because of invariance principles. In this chapter we
deal with problems whose exact solutions are not known. The methods
used are intuitive and cannot be regarded as proven. However, they
give a qualitative indication of where the solution lies and in many
cases have been checked by computer simulation. They should have
heuristic value and I hope that they may point the way to rigorous,
exact solutions at some later time.

11.1 Gene frequency distribution (global) in a structured population

One of main emphases in this section is to see how closely a
structured population behaves like a panmictic population, and to dis-
cover the parameters which determine the fate of genes.

Here we deal with selectively neutral genes which are maintained
by mutation and random drift, and will obtain formulae for the stationary distributions of gene frequencies for a population of the
general structure studied in the preceding chapter. Briefly, the
population consists of a finite number of colonies of arbitrary finite
size. Mating takes place independently in each colony and then migration occurs between generations. We assume that the population structure is fixed so that a steady state exists.

Diallelic case. Assume that there are two alleles A and a at a
locus under consideration, and that the mutation rate from A to a is u
and the reverse rate is v. Let N_i be the size of colony i ($2N_i$ genes) and

x_i be the gene frequency of A in colony i. Now let

$$X \equiv \frac{1}{N} \sum_i x_i N_i$$

where $N = \sum N_i$, the total population size. Therefore X is the mean gene frequency of the A allele in   the whole population. Thus X is a random variable and the change of this random variable forms a stochastic process. We assume the process can be approximated by a diffusion process as we did in section 10.4. By the assumption that each colony mates independent of others, the change in the value of X due to random sampling of gametes in one generation is the weighted mean of the change in each colony. Thus

$$\Delta X = \frac{1}{N} \sum_i \Delta x_i N_i$$

where Δx_i is the change in colony i. The mean change due to the random sampling of gametes is zero, i.e., $E\{\Delta x\} = 0$. The variance $(V_{\delta X})$ can be computed:

(11.1)
$$E\{(\Delta X)^2\} = \frac{1}{N^2} \sum_i \sum_j E\{\Delta x_i \Delta x_j\} N_i N_j$$

$$= \frac{1}{2N} \sum \frac{x_i(1 - x_i)N_i}{N}$$

as in (10.47). Now the change in the value of X due to mutation is

(11.2)
$$M_{\delta X} = (1 - u)X + v(1 - X) = -uX + v(1 - X).$$

Note that the quantity summed in (11.1) is proportional to the actual heterozygote probability, while X(1-X) is the corresponding probability for two genes chosen at random from the whole population. We will write

(11.3)
$$H(X) \equiv \frac{1}{N} \sum x_i(1 - x_i)N_i .$$

The ratio of H(X) to X(1-X) measures the degree of population structure in reference to a panmictic population of corresponding size. For a random mating population H(X) = X(1-X); therefore the ratio is one. If a population highly structured   and alleles are strongly localized, the value of H(X) will be much smaller than the value of X(1-X)

and thus the ratio will be small. Robertson (1964) and Maruyama
(1971b) have shown that the rate of decay of genetic variability measured by the probability X(1-X) is equal to

$$\frac{1}{2N} \frac{H(X)}{X(1-X)} \cdot$$

Noting that the rate for a random mating population is 1/2N, the ratio
H(X)/X(1-X) is the factor due to the population structure and therefore
is a measure of the structure.

Using the notation defined in (11.3), we can write the KBE as

(11.4) $$\frac{\partial u(t, X)}{\partial t} = \frac{H(X)}{4N} \frac{\partial^2 u(t, X)}{\partial x^2} + \{-uX + v(1 - X)\}\frac{\partial u(t, X)}{\partial X}$$

where

$$u(t, X) = \int_0^1 P(t, X, Y)f(Y)dY$$

with P(t, X, Y) = the transition probability density that the frequency
of A-gene is Y at time t given that it was X at time 0. From the KBE (11.4)
together with (7.50), the equilibrium distribution is

(11.5) $$\phi(Y) = \frac{C}{H(Y)} \exp\left\{\int_0^Y \frac{-4Nu\xi + 4Nv(1 - \xi)}{H(\xi)} d\xi\right\}$$

which is a special case of Wright's formula (7.50).

Formula (11.5) is not very informative, but we can change this to
a formula that reveals more about the effect of population structure on
the gene frequency distribution. Observe first that using the mean
value theorem of integral calculus

$$\int_0^Y \frac{\xi}{H(\xi)} d\xi = \int_0^Y \frac{\xi(1 - \xi)}{H(\xi)(1 - \xi)} d\xi$$

$$= \frac{Y_0(1 - Y_0)}{H(Y_0)} \int_0^Y \frac{d\xi}{1 - \xi}$$

$$= -\frac{Y_0(1 - Y_0)}{H(Y_0)} \log(1 - Y)$$

where Y_0 is a number between 0 and Y, i.e., $0 < Y_0 < Y$. Similarly

$$\int_0^Y \frac{1 - \xi}{H(\xi)} = \frac{Y_1(1 - Y_1)}{H(Y_1)} \log Y \ .$$

Thus

$$\int_0^Y \frac{-4Nu\xi + 4Nv(1 - \xi)}{H(\xi)} \, d\xi$$

$$= 4Nu \frac{Y_0(1 - Y_0)}{H(Y_0)} \log(1 - Y) + 4Nv \frac{Y_1(1 - Y_1)}{H(Y_1)} \log Y$$

where $0 < Y_0, Y_1 < Y$. Therefore formula (11.5) can be written as

$$(11.6) \qquad \phi(Y) = \frac{C}{H(Y)} \, Y^{4Nv \frac{Y_0(1-Y_0)}{H(Y_0)}} (1 - Y)^{4Nu \frac{Y_1(1-Y_1)}{H(Y_1)}}$$

in which C is a normalizing constant. If a population is panmictic, $Y(1-Y) = H(Y)$ for all Y, and the formula reduces to Wright's formula for the two allelic case of reversible mutation:

$$(11.7) \qquad \phi(Y) = CY^{4Nv-1}(1 - Y)^{4Nu-1} \ .$$

It seems neceasary to comment on the meaning of H(Y). Since we are dealing with an equilibrium state, H(Y) is the expectation, or the weighted mean, of $\sum_i y_i(1-y_i)N_i/N$ for all possible sample paths, with restriction $Y = \sum_i y_i N_i/N$. It is remarkable that formula (11.7) has essentially the same pattern as (11.6) in which the parameters 4Nv and 4Nu are multiplied by a factor $Y(1-Y)/H(Y)$. The effect of the population structure is therefore similar to having a larger population size. If the population structure prevents strongly the gene mixture and causes localization of alleles, then $Y(1-Y) \gg H(Y)$ and the ratio $Y(1-Y)/H(Y)$ would be large. In such a case the distribution of gene frequency would resemble that of a much larger population than the actual size. At any rate it is important that the population structure does not change the essential pattern of the gene frequency distribution.

Infinite allele case. Here we assume that the number of possible alleles at a locus under consideration is very large and thus every

mutant is new to the population. Every allele is selectively neutral
and mutation rate is u per gene per generation. This is the infinite
allele model of Kimura and Crow (1964), which is essentially the same
model studied by Wright (1948) who obtained a formula for the distribution of gene frequency. Kimura and Crow (1964) investigated the
number of alleles and also the distribution of alleles for this model.
Ewens (1963, 1969) studied the sojourn time and the number of alleles
maintained in an equilibrium population. After Kimura (1968) proposed the neutrality hypothesis of molecular evolution, this model has
played an important role in that field and has been studied extensively.
For a review of the model, see Nei (1975).

The population structure dealt here is the same as that in the
preceding sub-section.

We first consider a particular allele existing in the population
and ask the sojourn time for that allele at given gene frequency. Let
x_i be the frequency of that particular allele in colony i, and let

$$X = \frac{1}{N} \Sigma x_i N_i .$$

The change of X in time is a stochastistic process. As before we
assume it can be approximated by a diffusion process. Then by the
assumption that each colony mates independent of the others, the variance $(V_{\delta X})$ of the change in X in one generation is the sum of the
variances in each colony, that is, as before,

(11.8)
$$V_{\delta X} = \frac{1}{2N} \underset{i}{\Sigma} \frac{x_i(1 - x_i)N_i}{N} .$$

The mean change $(M_{\delta X})$ of X in one generation is simply

$$M_{\delta X} = (1 - u)X - X = -uX .$$

The KBE governing the process X is therefore

$$\frac{\partial P}{\partial t} = \frac{H(X)}{4N} \frac{\partial^2 P}{\partial X^2} - uX \frac{\partial P}{\partial X}$$

where $P = P(t, X, Y)$ is the transition density function, and $H(X)$ is
given by (11.3).

The average sojourn time of sample paths (Green's function) is
given by an appropriate solution of the differential equation

$$(11.9) \qquad \frac{H(X)}{4N} \frac{d^2\Phi(X,\ Y)}{dX^2} - uX \frac{d\Phi(X,\ Y)}{dX} + \delta(X - Y) = 0$$

where $\Phi(X,\ Y)dY$ is the average sojourn time that a path stays in a range $(Y,\ Y+dY)$ given that it starts from $(x_1,\ x_2,\ \cdots)$ for which $X = \sum_i x_i N_i/N$. The differential equation (11.9) can be integrated and the appropriate boundary conditions are

$$\Phi(0,\ Y) = 0$$

and

$$\frac{\partial\Phi(X,\ Y)}{X}\bigg|_{X=1} = 0\ ,$$

(see section 4.9). The solution satisfying these conditions is

$$(11.10)\quad \Phi(X,Y) = \frac{4N}{H(Y)} \exp\left\{4Nu \int_0^Y \frac{\xi}{H(\xi)} d\xi\right\} \int_0^X \exp\left\{4Nu \int_0^\theta \frac{\xi}{H(\xi)} d\xi\right\} d\theta$$

$$\text{for } X > Y,$$

$$= \frac{4N}{H(Y)} \exp\left\{4Nu \int_0^Y \frac{\xi}{H(\xi)} d\xi\right\} \int_0^Y \exp\left\{4Nu \int_0^\theta \frac{\xi}{H(\xi)} d\xi\right\} d\theta$$

$$\text{for } X < Y.$$

The most important case of formula (11.10) here would be for $X = 1/2N$, because this represents the initial condition for every allele which is assumed to start from a single mutant. For this case the solution is given by

$$(11.11) \qquad \Phi(Y) = \frac{4N}{H(Y)} \frac{1}{2N} \exp\left\{4Nu \int_0^Y \frac{\xi}{H(\xi)} d\xi\right\}.$$

Here we used an approximation that the integral from 0 to $1/2N$ on the exponential function is replaced by $1/2N$. If we apply the mean value theorem of integral calculus to $\int_0^Y \xi/H(\xi)d\xi$, we have

$$(11.12)\quad \Phi(Y) = \frac{4N}{H(Y)} \frac{1}{2N} \exp\left\{4Nu \frac{Y_0(1-Y_0)}{H(Y_0)} \log(1-Y)\right\} = \frac{2}{H(Y)}(1-Y)^{4Nu\frac{Y_0(1-Y_0)}{H(Y_0)}}.$$

Here $H(Y_0)$ is the average of $\Sigma y_i(1-y_i)N_i/N$ with $Y_0 = y_iN_i/N$ for the paths starting from identical initial conditions. As before, the formula involves one indeterminable factor $Y_0(1-Y_0)/H(Y_0)$ which is the ratio of the probability that two randomly chosen homologous genes from the whole population are not identical to the corresponding probability for two genes from a single colony. If the population is panmictic, $Y(1-Y) = H(Y)$ for all Y and the formula reduces to

$$(11.13) \qquad 2Y^{-1}(1 - Y)^{4Nu-1}$$

which is the same as (4.48).

From formula (11.12), we can obtain the expected number of alleles with frequency Y in an equilibrium population. Since (11.12) is the sojourn time, to obtain the distribution of the number of alleles we multiply (11.12) by the expected number of new mutant alleles to be introduced in each generation. Here we are applying ergodic theory which asserts that a time average is equal to the corresponding space average. Then the expected number of alleles with frequency Y is given by

$$(11.14) \qquad \Phi(Y) = \frac{4Nu}{H(Y)}(1 - Y)^{4Nu\frac{Y_0(1-Y_0)}{H(Y_0)}}$$

where $0 < Y_0 < Y$ and $H(Y_0)$ is the expectation of $\Sigma y_i(1-y_i)N_i/N$ with $Y_0 = \Sigma y_iN_i/N$. If the population is panmictic the formula reduces to

$$\Phi(Y) = 4NuY^{-1}(1 - Y)^{4Nu-1}$$

which is the same as (4.49). Formula (11.14) is the density of the number of alleles whose frequency is Y.

It is interesting to note that if we are to measure the quantity $H(Y)$ which can be interpreted as the actual heterozygote frequency, we multiply $\Phi(Y)$ of (11.6) by $H(Y)$ and obtain

$$(11.15) \qquad \psi(Y) = 4Nu(1 - Y)^{4Nu\frac{Y_0(1-Y_0)}{H(Y_0)}}.$$

In the case of a panmictic population

$$(11.16) \qquad \psi(Y) = 4Nu(1 - Y)^{4Nu}.$$

Therefore in terms of the actual heterozygotes, the pattern of distribution is exactly the same. The influence of the population structure is to increase the effective size of population by a factor of $Y_0(1 - Y_0)/H(Y_0)$.

We may inquire into the effect of population structure on the average homozygote (or heterozygote) probability, that is the probability that two randomly chosen genes from a single colony are identical (not identical) by descent. The average heterozygote probability ($\bar{H}_0$) can be calculated by integrating $\Phi(Y)$ of (11.15) or (11.16), i.e.,

$$(11.17) \qquad \bar{H}_0 = \int_0^1 \psi(Y) dY \ ,$$

$$(11.18) \qquad \bar{H}_0 = \frac{4Nu}{1 + 4Nu} \qquad \text{for panmictic case,}$$

and

$$(11.19) \qquad \bar{H}_0 = \frac{4Nu}{1 + \dfrac{4NuY_0(1-Y_0)}{H(Y_0)}} \qquad \text{for structured case,}$$

where Y_0 is a number between 0 and 1. The formula for a panmictic case has been known, (Malécot, 1948; Kimura and Crow, 1964).

We now let the global genetic variability ($\bar{H}$) be the probability that two randomly chosen genes from the whole population are not identical by descent. It can be calculated by

$$(11.20) \qquad \bar{H} = \int_0^1 Y(1 - Y)\psi(y) dY$$

where $\psi(Y)$ is the gene frequency distribution given in (11.14). For a case of a panmictic population, $\bar{H}_0 = \bar{H}$. For a structured population, it is known that

$$(11.21) \qquad \bar{H} = 1 - \frac{\bar{H}_0}{4Nu} \ .$$

And this relationship between $\bar{H}$ and $\bar{H}_0$ is independent of the population structure, (Crow and Maruyama, 1971). If $\bar{H} = \bar{H}_0$, then formula (11.21) reduces to (11.18).

Now compare $\bar{H}_0$ of (11.18) and (11.19). For a given value of $4Nu$, the larger the ratio $Y_0(1 - Y_0)/H(Y_0)$, the smaller is value of $\bar{H}_0$, because the denominator of the right hand side of (11.19)

increases as the ratio increases. Therefore if the population structure causes localization of alleles, the local heterozygosity, $\bar{H}_0$, is
decreased. On the other hand, formula (11.21) asserts that as the
local heterozygosity ($\bar{H}_0$) decreases, the global heterozygosity ($\bar{H}$)
increases. And if the ratio $Y_0(1-Y_0)/H(Y_0)$ is greater than unity, $\bar{H}$
is greater than $4Nu/(1+4Nu)$ which is the genetic variability for a
panmictic population. A more exact relationship can be obtained by
substituting $\bar{H}_0$ of (11.19) into (11.21). The result is

$$(11.22) \qquad \bar{H} = \frac{4Nu \dfrac{Y_0(1 - Y_0)}{H(Y_0)}}{1 + 4Nu \dfrac{Y_0(1 - Y_0)}{H(Y_0)}} \ .$$

If $Y_0(1-Y_0) = H(Y_0)$, the above formula reduces to formula (11.18)
which is for a panmictic population. Throughout this section the ratio $Y(1-Y)/H(Y)$ has played an important role.

11.2 Distribution of local gene frequencies

The distribution of gene frequencies in a single colony of the
structured population appears to be a very difficult problem. This
question is probably much more difficult than that of the global distribution. Maruyama (1972b) has made an attempt to attack this problem
for a stepping stone model. Although Maruyama's formula is incorrect,
there are some indication that the distribution for selectively neutral alleles is approximately a beta function

$$(11.23) \qquad \phi(y) = cy^{\alpha-1}(1 - y)^{\beta-1}$$

where α and β are related to the mean and the variance of x. Namely

$$\bar{y} = \int_0^1 y\phi(y)\,dy$$

and

$$\sigma_y^2 = \int_0^1 (y - \bar{y})^2\phi(y)\,dy \ .$$

There has been no mathematical justification for formula (11.23).
Nevertheless the formula appears to be a fairly good approximation.
For the infinite allele case, the equivalent formula is

$$(11.24) \qquad \Phi(y) = \beta y^{-1}(1 - y)^{\beta - 1}$$

in which

$$(11.25) \qquad \beta = (1 - f_0)/f_0 \ ,$$

where f_0 is the average homozygosity probability in the colony under consideration. In order to check these formula I have performed a rather extensive simulation. The results of simulations indicate that these formulae are not exact, but are good approximations. Therefore the distribution is mainly determined a single parameter, i.e., the average homozygote probability (f_0). Tables 11.1 and 2 are two examples of the simulation results.

Table 11.1 presents a case of a population consisting of one large subpopulation (N = 200) along with ten small subpopulations (N = 10 each). At every generation, each small subpopulation exchanges one individual with the large subpopulation. Every mutant is new (infinite allele model) and the mutation rate per gene per generation is u = 0.003. In the table, the simulation result is compared with the theoretical expectation based on formula (11.24) in which the parameter β is calculated by substituting the mean homozygote probability of the simulation into (11.25).

Table 11.1 Comparison of simulation result to
theoretical expectation based on formula (11.24).

Small (single) subpopulation

Gene frequency	0-.1	.1-.2	.2-.3	.3-.4	.4-.5	.5-1.0	
Number of alleles	2.729	1.375	.698	.392	.221	.199	(sim.)
	3.218	1.065	.561	.335	.208	.264	(theor.)

Large subpopulation

Gene frequency	0-.01	.01-.02	.02-.05	.05-.10	.10-.20	.20-.30	
Number of alleles	7.068	2.290	2.888	2.014	1.777	.736	(sim.)
	7.713	2.280	3.001	2.071	1.675	.691	(theor.)

Gene frequency	.3-.4	.4-.5	.5-.6	.6-.7	.8-.9	.9-1.0	
Number of alleles	.321	.170	.070	.022	.016	0	(sim.)
	.330	.161	.075	.032	.011	.002	(theor.)

Table 11.2 presents a case of population consisting of ten circularly arranged colonies (N = 10 each). Migration rate between adjacent subpopulations is 0.1 (mN = 1) and the mutation rate is u = 0.0005. For this case, we can calculate the theoretical expectation of the local homozygote probability, using formula (9.19).

Table 11.2 Comparison of simulation result with
the theoretical expactation.

Gene frequency	0-.1	.1-.2	.2-.3	.3-.4	.4-.5	.5-.6	
Number of alleles	.199	.104	.089	.080	.078	.075	(sim.)
	.283	.121	.085	.080	.063	.063	(theor.)

Gene frequency	.6-.7	.7-.8	.8-.9	.9-1.0	
Number of alleles	.075	.083	.098	.072	(sim.)
	.066	.075	.091	.066	(theor.)

The average homozygote probability
0.849 (sim.), 0.843 (theor.)

From these simulations and also others, it appears to be that the number of low frequency alleles is substantially fewer than the expectation based on formulae (11.23) or (11.24). But for the other frequency classes of alleles, the numbers are in a fairly good agreement. It is, therefore, quite remarkable that a quantity, the average homozygote probability, can determine much of the behavior of the whole frequency distribution. It will be shown later that the average homozygote probability also determines the variance of the homozygosity in a single colony and the variance of the probability of identity of two randomly chosen genes which are separated by a given distance.

11.3 Random drift in a structured population

Here we shall consider the effect of the population structure on the speed of gene substitution in a population. If we ignore the mutation pressure in KBE (11.4), the equation becomes

$$(11.26) \qquad \frac{\partial u(t, X)}{\partial t} = \frac{H(X)}{4N} \frac{\partial^2 u(t, X)}{\partial X^2}$$

where $H(X) = \Sigma\, x_i (1 - x_i) N_i / N$. Thus if we denote by $T(X)$ the time to either fixation or extinction of an allele in question, as a function of the initial gene frequency, then by equation (4.6), T(X) satisfies

(11.27) $$\frac{H(X)}{4N} \frac{d^2 T(X)}{dx^2} + 1 = 0.$$

The above equation cannot be integrated in a simple form. Therefore, instead of solving (11.27) directly, we obtain Green's function for the process which satisfies

(11.28) $$\frac{H(X)}{4N} \frac{d^2 \Phi(X, Y)}{dx^2} + \delta(X - Y) = 0$$

with boundary conditions

$$\Phi(0, Y) = \Phi(1, Y) = 0.$$

The appropriate solution is

(11.29) $$\Phi(X, Y) = \frac{4N(1 - Y)X}{H(Y)} \quad \text{for } X < Y,$$

$$= \frac{4N(1 - Y)X}{H(Y)} - \frac{4N(X - Y)}{H(Y)} \quad \text{for } X > Y.$$

The panmictic analogue of this Green's function is given in (4.23).

To obtain the time to fixation or extinction, we integrate $\Phi(X,Y)$ and obtain

(11.30) $$T(x) = \int_0^1 \Phi(X, Y)dY$$

$$= 4NX \int_X^1 \frac{1 - Y}{H(Y)} dY + 4N \int_0^X \frac{X(1 - Y)-(X - Y)}{H(Y)} dY$$

$$= 4NX \int_X^1 \frac{1 - Y}{H(Y)} dY + 4N(1 - X) \int_0^X \frac{Y}{H(Y)} dY.$$

In order to express this integral in a more appealing form, we use the mean value theorem of integral calculus,

(11.31) $$\int_X^1 \frac{1 - Y}{H(Y)} dY = \int_X^1 \frac{(1 - Y)Y}{YH(Y)} dY = - \frac{Y_1(1 - Y_1)}{H(Y_1)} \log X$$

and

$$(11.32) \quad \int_0^X \frac{Y}{H(Y)} \, dY = \int_0^X \frac{(1-Y)Y}{(1-Y)H(Y)} \, dY = -\frac{Y_0(1-Y_0)}{H(Y_0)} \log(1-X)$$

where $0 < Y_0 < X$ and $X < Y_1 < 1$. Substitution of (11.31) and (11.32) into (11.30), we have

$$(11.33) \quad T(X) = -4N \left\{ \frac{Y_0(1-Y_0)}{H(Y_0)}(1-X)\log(1-X) + \frac{Y_1(1-Y_1)}{H(Y_1)} X \log X \right\}.$$

As for the equilibrium distribution of gene frequencies (11.6) and (11.14), the ratio $Y(1-Y)/H(Y)$ determines the effect of the population structure. Hence if the ratio stays always near unity, the effect is small and the time to fixation or extinction is almost the same as that of a panmictic population of the same size. But the ratio can be large in certain cases. For instance, if two populations of different genetic make-up come into contact with low migration rate and if one of the alleles is to spread over the whole population, it may take a considerably longer time than it would take for a panmictic population. For a situation like this, we can use formula (11.33) to make an upper bound on the fixation or extinction time. For instance, suppose that one of the two subpopulations has A and a alleles in 1 : 9 ratio, while the other subpopulation has A and a alleles in 9 : 1 ratio. Then the ratio when they come into contact is

$$\frac{0.5(1-0.5)}{\frac{1}{2}\{0.1(1-0.1)+0.9(1-0.1)\}} = \frac{.25}{.09} = 2.78 .$$

Hence unless migration between the two subpopulations is so low that local random drift predominates, the time will not be greater than about 2.8.

We shall next investigate the time to fixation for the selected sample paths destined to go to fixation. If we let $\Phi_1(X, Y)$ be Green's function for such sample paths and $G(X, Y) = X\Phi_1(X, Y)$, then noting (5.5) and $u(X) = X$

$$(11.34) \quad \frac{H(X)}{4N} \frac{d^2G(X, Y)}{dX^2} + X\delta(X - Y) = 0$$

with boundary conditions $G(0, Y) = G(1, Y) = 0$. The solution is

(11.35) $\Phi_1(X, Y) = \frac{1}{X} G(X, Y) = 4N \dfrac{Y(1 - Y)}{H(Y)}$ for $X < Y$,

$= 4N \dfrac{Y^2(1 - X)}{XH(Y)}$ for $X > Y$.

If $X \approx 0$, Green's function is independent of X for all Y, and it deviates from that of the panmictic case by a factor of $Y(1-Y)/H(Y)$, (cf. 5.18). The time required for a gene substitution can be calculated by

(11.36) $T_1(X) = \displaystyle\int_0^1 \Phi_1(X, Y)\,dY$

$= 4N \displaystyle\int_0^X \dfrac{Y^2(1 - X)}{XH(Y)}\,dY + 4N \int_X^1 \dfrac{Y(1 - Y)}{H(Y)}\,dY$

$= -4N \dfrac{(1-X)}{X} \dfrac{Y_0(1-Y_0)}{H(Y_0)} [X + \log(1-X)]$

$+ 4N(1-X) \dfrac{Y_1(1-Y_1)}{H(Y_1)}$

where $0 < Y_0 < X$ and $X < Y_1 < 1$. Therefore if $X = 1/2N$, we have

(11.37) $T_1(\frac{1}{2N}) = 4N \dfrac{Y_1(1 - Y_1)}{H(Y_1)}$

where $0 < Y_1 < 1$. Noting that the fixation time for a panmictic case is 4N, the deviation due to the population structure is the factor $Y_1(1-Y_1)/H(Y_1)$. In formula (11.36) if we let $X = 1-1/2N$, then

(11.38) $T_1(1 - \frac{1}{2N}) = 2 \dfrac{Y_0(1 - Y_0)}{H(Y_0)} \log 2N$

where $0 < Y_0 < 1/2N$. Formula (11.38) can be viewed as the time for a singly present gene to be lost from the population without going to fixation.

CHAPTER 12

SOME SPECIAL PROBLEMS

12.1 Variance of homozygote probability for the infinite neutral allele model

We have shown that the mean homozygote probability ($\bar{f}$) for this model is equal to $1/(1+4Nu)$ at equilibrium. It is evident that this probability varies in time and among loci. Therefore the distribution or the variance of the homozygote probability is of considerable importance. Following Stewart (1976) we shall derive the variance.

To do this let us look at a problem of multiple alleles. Assume that there are $k+1$ alleles A_0, A_1, $\cdots$ A_k at a locus and the mutation rate from one allele to another is u/k. We assume also that all the alleles are selectively neutral. Let x_i be the frequency of allele A_i ($\sum_{i=0}^{k} x_i = 1$), and $\phi(t; x_i, \cdots, x_k; y_1, \cdots, y_k)$ be the transitional probability density which is a k-dimensional analogue of those discussed before. Then the KBE for this density is

$$(12.1) \quad \frac{\partial \phi}{\partial t} = \frac{1}{2} \sum_{i=1}^{k} V_{\delta x_i} \frac{\partial^2 \phi}{\partial x_i^2} + \frac{1}{2} \sum_{\substack{i=1 \\ (i \neq j)}}^{k} \sum_{j=1}^{k} C_{\delta x_i \delta x_j} \frac{\partial^2}{\partial x_i \partial x_j} + \sum_{i=1}^{k} M_{\delta x_i} \frac{\partial \phi}{\partial x_i}$$

where

$V_{\delta x_i}$ = the infinitesimal variance of x_i, which can be approximated by the sampling variance, $x_i(1-x_i)/2N$,

$C_{\delta x_i \delta x_j}$ = the infinitesimal covariance between x_i and x_j, approximated by the sampling covariance, $-x_i x_j/2N$,

$M_{\delta x_i}$ = the infinitesimal mean change,
$(1-u)x_i + (1-x_i)\frac{u}{k} - x = \frac{u}{k} - (\frac{k+1}{k})ux_i$

Therefore the KBE (12.1) becomes

$$(12.2) \qquad \frac{\partial \phi}{\partial t} = \frac{1}{2} \sum_{i=1}^{k} \frac{x_i(1-x_i)}{2N} \frac{\partial^2 \phi}{\partial x_i^2} - \frac{1}{2} \sum_{i=1}^{k} \sum_{\substack{j=1 \\ (i \neq j)}}^{k} \frac{x_i x_j}{2N} \frac{\partial^2 \phi}{\partial x_i \partial x_j}$$

$$+ \sum_{i=1}^{k} \left\{ \frac{u}{k} - \frac{k+1}{k} u x_i \right\} \frac{\partial \phi}{\partial x_i} .$$

The eigenfunction of (12.2), associated with eigenvalue 0 is given by

$$(12.3) \qquad \phi(x_1, x_2, \cdots x_k) = \frac{\Gamma(4Nu)}{\Gamma(\frac{4Nu}{k})^k} x_1^{\frac{4Nu}{k} - 1} x_2^{\frac{4Nu}{k} - 1} \cdots$$

$$x_k^{\frac{4Nu}{k} - 1} (1-x_1-x_2- \cdots -x_k)^{\frac{4Nu}{k} - 1}$$

where $\Gamma(\cdot)$ is the Gamma function. Formula (12.3) gives the probability density that the frequency of allele A_1 is x_1, the frequency of A_2 is x_2 and so on, and it can be rewritten as

$$\phi(x_1, x_2, \cdots x_k) = \frac{\Gamma(4Nu)}{\Gamma(\frac{4Nu}{k})^k} (x_0 \cdot x_1 \cdots x_k)^{\frac{4Nu}{k} - 1}$$

where $x_0 = 1-x_1-x_2- \cdots -x_k$. The following calculations are known as Dirichlet's integral and give the moments for the homozygote probability (f).

$$(12.4) \qquad \mu_n' = \int_0^1 \int_0^{1-x_1} \cdots$$

$$\int_0^{1-x_1-x_2- \cdots -x_{k-1}} (x_0^2+x_1^2+ \cdots +x_k^2)^n \phi(x_1, x_2, \cdots, x_k) dx_k \cdots dx_2 dx_1$$

$$= \int_0^1 \int_0^{1-x_1} \cdots \int_0^{1-x_1- \cdots -x_{k-1}} \sum_{n_0+ \cdots +n_k=k} \frac{n!}{n_0! n_1! \cdots n_k!}$$

$$\times x_0^{2n_0} x_1^{2n_1} \cdots x_k^{2n_k} \frac{\Gamma(4Nu)}{\Gamma(\frac{4Nu}{k})^k} (x_0 \cdot x_1 \cdots x_k)^{\frac{4Nu}{k}-1} dx_k \cdots dx_1$$

$$
= \sum_{n_0 + \cdots + n_k = n} \frac{n!}{n_0! n_1! \cdots n_k!} \frac{\Gamma(4Nu)}{\Gamma(\frac{4Nu}{k})^k} \frac{\Gamma(2n_0 + \frac{4Nu}{k}) \cdots \Gamma(2n_k + \frac{4Nu}{k})}{\Gamma\left(\sum_{i=0}^{k}(2n_i + \frac{4Nu}{k})\right)}
$$

$$
= \frac{n! \Gamma(4Nu)}{\Gamma(4Nu + 2k)} \sum_{n_0 + \cdots + n_k = n} \prod_{i=0}^{k} \frac{\Gamma(\frac{4Nu}{k} + 2n_i)}{n_i! \Gamma(\frac{4Nu}{k})}
$$

$$
= \frac{k!}{\Gamma(4Nu + 2k)} \sum_{\nu=1}^{n} \frac{(k+1)!}{\nu!(k+1-\nu)!} \sum_{\substack{m_1 + \cdots + m_\nu = n \\ m_j \geq 1}} \prod_{j=1}^{\nu} \frac{\Gamma(\frac{4Nu}{k} + 2m_j)}{m_j! \Gamma(\frac{4Nu}{k})} .
$$

When n = 1, (12.4) becomes

$$
\mu_1' = E(f) = \frac{(k+1)\frac{4Nu}{k}(\frac{4Nu}{k} + 1)}{\Gamma\left(\frac{4Nu(k+1)}{k} + 1\right)} = \frac{\frac{\theta}{k} + 1}{\frac{(k+1)\theta}{k} + 1}
$$

which is the same as obtained by Kimura (1968a) for k+1 alleles. When n = 2, (12.4) yields

$$
\mu_2' = \frac{2\Gamma\left((k+1)\frac{4Nu}{k}\right)}{\Gamma\left(\frac{(k+1)4Nu}{k} + 4\right)} \left\{ (k+1) \frac{\Gamma(\frac{4Nu}{k} + 4)}{2\Gamma(\frac{4Nu}{k})} + \frac{k(k+1)}{2} \left(\frac{\Gamma(\frac{4Nu}{k} + 2)}{\Gamma(\frac{4Nu}{k})}\right)^2 \right\}
$$

$$
= \frac{(\frac{4Nu}{k} + 1)\left\{(\frac{4Nu}{k} + 2)(\frac{4Nu}{k} + 3) + k\frac{4Nu}{k}(\frac{4Nu}{k} + 1)\right\}}{\left(\frac{(k+1)4Nu}{k} + 1\right)\left(\frac{(k+1)4Nu}{k} + 2\right)\left(\frac{(k+1)4Nu}{k} + 3\right)} .
$$

The variance of f is

$$
\sigma_f^2 = \mu_2' - (\mu_1')^2 = \frac{2k \frac{4Nu}{k}(\frac{4Nu}{k} + 1)}{\left(\frac{(k+1)4Nu}{k} + 1\right)^2\left(\frac{(k+1)4Nu}{k} + 2\right)\left(\frac{(k+1)4Nu}{k} + 3\right)}
$$

$$
= \frac{8Nu(1 + \frac{4Nu}{k})}{(1 + 4Nu + \frac{4Nu}{k})(2 + 4Nu + \frac{4Nu}{k})(3 + 4Nu + \frac{4Nu}{k})} .
$$

As k → ∞ (the infinite allele model),

$$
(12.5) \qquad \sigma_f^2 = \frac{2\theta}{(1 + \theta)^2(2 + \theta)(3 + \theta)}
$$

where θ = 4Nu, (Stewart, 1976). See also Li and Nei (1975).

12.2 Variance of homozygote probability in a geographically structured population

In chapter 9, we studied the probability of gene identity as a function of geographical distance, and it was possible to determine explicitly the formulae for some special cases of the structure. The formulae obtained in chapter 9 are means. As discussed in the preceding section, the actual probability fluctuates around the mean. Exact solutions which are analogues of formula (12.5) are not known. Like the gene frequency distribution in a structured population, it is conceivable that the amount of the fluctuation is very much determined by the mean homozygote probability and that the whole population and any local population behave very much like a panmictic population having the same mean homozygosity. Indeed, some computer simulations have confirmed this conjecture.

An approximation of the variance of the global homozygosity which is the probability of identity for two randomly chosen genes from the whole population is given by formula (12.5) in which

$$\theta = \frac{1 - \bar{f}_0}{\bar{f}}$$

where $\bar{f}_0$ is the average of the local homozygosities and $\bar{f}$ is the global homozygosity. An approximation of the local variance of homozygosity is also given by formula (12.5) in which

(12.6)
$$\theta = \frac{1 - f_0}{f_0}$$

where f_0 is the mean homozygosity at a particular locality under consideration. The variance of the identity probability for two genes separated by a given distance can be approximated by

(12.7)
$$\sigma^2_{ij} = \sqrt{\sigma^2_{ii} \sigma^2_{jj}} \; \frac{f_{ij}}{\sqrt{f_{ii} f_{jj}}}$$

in which σ^2_{ii} and σ^2_{jj} are the variances of the gene identity at localities i and j respectively, and f_{ii}, f_{jj} and f_{ij} are the mean identity probabilities for two genes from locality i, from locality j and from i and j respectively.

Fig. 12.1 (a) and (b) compare these theoretical approximations with simulation results. The circular stepping stone structure was used for the simulations, because it is possible to determine f_0, f_i and $\bar{f}$ theoretically.

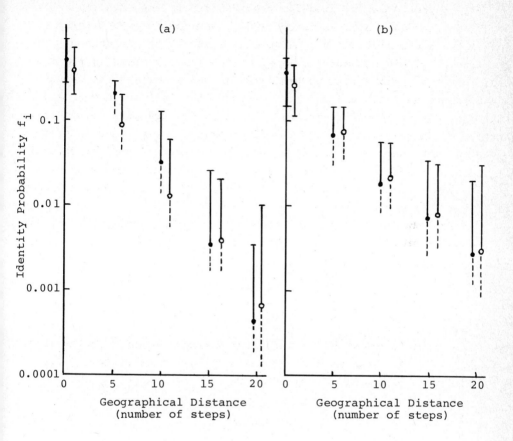

Fig. 12.1 The mean and variance of the probability that two homologous genes separated by a given distance are identical by descent. Circular stepping stone structure was used. Circles in the graphs indicate the mean probabilities; solid = simulation result and empty = theoretical. Vertical lines with bars indicate the corresponding standard deviations. The theoretical values were obtained from formula (12.5) with θ given in (12.6), and from (12.7). In (a), the number of colonies = 40, the colony size = 10 migration (restricted to adjacent colonies) rate = 0.4 (mN = 4), and mutation (the infinite allele model) rate = 0.01. In (b), the number of colonies and mutation rate are the same as in (a), the colony size = 5 and migration rate = 0.2 (mN = 1).

12.3 Number of alleles

Along with the heterozygote (or homozygote) probability, the number of different alleles found in a population or in a sample of genes is also of a quantity of interest. Ewens (1972) studied this problem for the infinite allele model.

First recall that the number of alleles whose frequencies are in (x, x+dx) is given by

$$\Phi(x)\,dx = 4Nux^{-1}(1 - x)^{4Nu-1}dx \ .$$

This is the frequency spectrum in reference to the whole population. Assume that we take a sample of n individuals (2n genes) from the population. Note that for any integer $m(m \leq 2n)$, the probability that m genes drawn at random are all of the same allelic type is

$$(12.8) \qquad \int_0^1 x^m \Phi(x)\,dx = (m-1)!/\{(1+\theta)(2+\theta) \cdots (m-1+\theta)\}$$

with $\theta = 4Nu$. This is because we are to pick up m genes which are all of the same type and of frequency x. Now consider these m genes as being drawn one by one from the population. Then from (12.8), the probability that all m genes are of the same allelic type, given that the first m-1 genes are, is

$$[(m-1)!/\{(1+\theta)(2+\theta)\cdots(m-1+\theta)\}] \div [(m-2)!/\{(1+\theta)(2+\theta)\cdots(m-2+\theta)\}]$$

$$= (m-1)/(m-1+\theta) \ .$$

Therefore, the probability that the m-th gene drawn is of different allelic type, given that the first m-1 genes drawn were all of the same allelic type, is

$$(12.9) \qquad 1 - \frac{m-1}{m-1+\theta} = \frac{\theta}{m-1+\theta} \ .$$

Now assume that no condition is made concerning the outcome of the first m-1 genes. Then the probability that the m-th gene is an allele not observed among the first m-1 draws is

$$\int_0^1 (1 - x)^{m-1}x\Phi(x)\,dx = \theta \int_0^1 (1 - x)^{m-1}x\{x^{-1}(1 - x)^{\theta-1}\}dx$$

$$= \theta/(\theta+m-1),$$

which is identical to (12.9). Interestingly, the unconditional probability that a new allele is picked up on the m-th draw is the same as
the conditional probability that the m-th gene is a new allele, given
that the first m-1 genes are all of the same type. This leads one to
speculate that a more general results is true, namely that the probability that the m-th gene is a new allele is given by (12.9), irrespective of any condition made concerning the number and frequencies
of the alleles observed for the first m-1 draws. This was proved to
be indeed true by Karlin and McGregor (1972).

Let n_i be the number of genes which belong to allele i and let k
be the number of different alleles observed in a sample of 2n genes,
$(\sum_{i=1}^{k} n_i = 2n)$. Using Karlin and McGregor's general result, Ewens proved
that the conditional vector $(n_1, n_2, \cdots, n_k)$ for a given k is independent of θ. This is equivalent to saying that k is a sufficient statis-
tic for $\theta = 4Nu$, (Ewens, 1972). Ewens showed also that the mean of k
can be calculated as follows:

$$(12.10) \qquad \bar{k} = \int_0^1 \{1 - (1-x)^{2n}\} \phi(x) dx$$

$$= \theta \int_0^1 \{1 - (1-x)^{2n}\} x^{-1} (1 - x)^{\theta-1} dx$$

$$= 1 + \frac{\theta}{1 + \theta} + \frac{\theta}{2 + \theta} + \cdots + \frac{\theta}{\theta + 2n - 1} .$$

The variance of k is given by

$$(12.11) \qquad Var(k) = \bar{k} - \left[1 + \frac{\theta^2}{(1 + \theta)^2} + \cdots + \frac{\theta^2}{(2n - 1 + \theta)^2} \right] .$$

12.4 Some properties of the stepwise mutation model

Instead of assuming every mutant is unique, Ohta and Kimura (1973)
have developed a model which is intended to account for electrophoretically detectable alleles. That is that alleles detected by
this method reflect the charge difference between the products of
alleles, and that, if mutation changes the charge, it does so by a fixed
amount in the positive or negative direction. This model can be illustrat-
ed schematically.

208

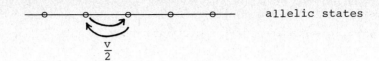

Fig. 12.1 Diagram illustrating the stepwise mutation model.

For this model, Ohta and Kimura (1973) obtained the probability that two randomly chosen genes are alleles of k-steps apart, k = 0, ±1, ±2, ··· . Here we shall derive Ohta and Kimura's result and some further properties of the model, (see also Wehrhahn, 1975).

Let $C_k(t)$ be the probability that two randomly chosen genes are alleles of k-steps apart at some time t. Consider $C_k(t+\Delta t)$ in terms of $C_k(t)$. In order for two genes to be of k-steps apart at time t+Δt, they have to be either k-steps, (k-1)-steps or (k+1)-steps apart at time t and

$$C_k(t+\Delta t) = (1-2v)C_k(t) + vC_{k+1}(t) + vC_{k-1}(t) + o(\Delta t).$$

This equation does not include the possibility of two genes coming from an identical gene at time t. This possibility is Δt/2N+o(Δt). If they come from a single gene, it contributes to $C_0(t+\Delta t)$. Therefore correcting the above equation for this term, we have, if k ≠ 0,

$$C_k(t+\Delta t) = (1-2v-\frac{1}{2N})C_k(t) + vC_{k+1}(t) + vC_{k-1}(t) + o(\Delta t)$$

$$+ \text{ higher order terms of v and 1/2N,}$$

and if k = 0

$$C_0(t+\Delta t) = (1-2v-\frac{1}{2N})C_0(t) + vC_{-1}(t) + vC_1(t) + o(\Delta t).$$

From these equations, we can easily derive

(12.12)
$$\begin{cases} \frac{dC_0(t)}{dt} = vC_{-1}(t) - (2v + \frac{1}{2N})C_0(t) + vC_1(t) + \frac{1}{2N}, \\ \frac{dC_k(t)}{dt} = vC_{k-1}(t) - (2v + \frac{1}{2N})C_k(t) + vC_{k+1}(t) \quad \text{for } k \neq 0. \end{cases}$$

This set of equations was given by Ohta and Kimura, and plays the basic role in determining the properties of the $C_k(t)$. If we let

$$A = \begin{pmatrix} -\alpha, & \beta & & & \\ \beta, & -\alpha, & \beta & & \text{\Large 0} \\ & \cdot & \cdot & \cdot & \\ & & \cdot & \cdot & \cdot \\ \text{\Large 0} & & & \beta, & -\alpha \end{pmatrix}$$
(All unspecified elements are zeros.)

where $\alpha = 2v + 1/2N$ and $\beta = v$, and J is a column vector $(\cdots, 0, 0, 1/2N, 0, 0, \cdots)*$. The system of equations given in (12.12) can be rewritten as

(12.13)
$$\frac{dC(t)}{dt} = AC(t) + J$$

where $C(t)$ is the column vector $(\cdots, C_{-1}(t), C_0(t), C_1(t), C_2(t), \cdots)*$. This system of differential equations can be solved formally System (12.13) can be written as

$$\frac{dC(t)}{dt} - AC(t) = J.$$

If we multiply both sides by e^{-tA} and integrate once, we have

$$e^{-At}C(t) = \left(\int_0^t e^{-A\xi}d\xi\right)J + B$$

where B is a constant matrix. For $t = 0$, the right side of the above equation must be equal to $C(0)$, and thus

$$C(0) = B.$$

Therefore

$$C(t) = e^{At}\left\{\left(\int_0^t e^{-A\xi}d\xi\right)J + C(0)\right\}.$$

Since

$$\int_0^t e^{-A\xi}d\xi = -A^{-1}e^{-A\xi}\Big|_0^t = A^{-1}(I - e^{-At}),$$

$$C(t) = e^{At}\left\{A^{-1}(I - e^{-At})J + C(0)\right\} = e^{At}A^{-1}J - A^{-1}J + e^{At}C(0).$$

Therefore symbolically the solution of (12.13) is

$$(12.14) \qquad C(t) = e^{At}\{A^{-1}J + C(0)\} - A^{-1}J$$

where A^{-1} is the inverse matrix of A. Let us continue a little more symbolic calculation. Assume that matrix A can be diagonalized

$$A = L\Lambda R$$

with

$$\Lambda = \begin{pmatrix} \ddots & & & & & 0 \\ & \lambda_{-1} & & & \\ & & \lambda_0 & & \\ & & & \lambda_1 & \\ 0 & & & & \ddots \end{pmatrix} ,$$

$$LR = I .$$

Then

$$e^{At} = I + At + \frac{t^2}{2!} A^2 + \cdots = L\left\{ I + t\Lambda + \frac{t^2}{2!}\Lambda^2 + \cdots \right\} R$$

$$= L \begin{pmatrix} \ddots & & & & 0 \\ & e^{\lambda_{-1}t} & & & \\ & & e^{\lambda_0 t} & & \\ & & & e^{\lambda_1 t} & \\ 0 & & & & \ddots \end{pmatrix} R .$$

Therefore if all the eigenvalues (more precisely the real parts of all the eigenvalues) are negative, $e^{\lambda_i t} \to 0$ for all i, and then

$$(12.15) \qquad e^{At} \to 0 \quad \text{(zero matrix)} .$$

And furthermore the solution given in (12.14) converges to

$$(12.16) \qquad \lim_{t \to \infty} C(t) = -A^{-1}J .$$

Now the problem is to determine the eigenvalues and eigenvectors of matrix A. To do this let us first assume that the matrix is of finite order and of the following form

$$\begin{pmatrix} -\alpha & \beta & & & & \beta \\ \beta & \cdot & \cdot & \cdot & \mathbf{0} & \\ & \cdot & \cdot & \cdot & & \\ & & \cdot & \cdot & \cdot & \\ & \mathbf{0} & & \cdot & \cdot & \beta \\ \beta & & & \beta & -\alpha \end{pmatrix}.$$

This matrix has been discussed in chapter 1 and corresponds to a closed circular space. The eigenvalues are

$$(12.17) \qquad \lambda_i = -\alpha + 2\beta \cos \frac{2\pi i}{n} = -\left\{ \frac{1}{2N} + 2v(1 - \cos \frac{2\pi i}{n}) \right\},$$

$$i = 0, 1, 2, \cdots, n-1 .$$

where n is the number of points on the circle. And the associated eigenvectors are column vectors $E_i \equiv (1, \cos \frac{2\pi i}{n}, \cos \frac{4\pi i}{n}, \cdots, \cos \frac{2(n-1)\pi i}{n})*$. From (12.16), we can say that the eigenvalues are all negative and therefore (12.15) holds. It is interesting to note that the largest two eigenvalues are

$$\lambda_0 = -\frac{1}{2N}$$

and

$$\lambda_1 = -\left\{ \frac{1}{2N} + 2v(1 - \cos \frac{2\pi}{n}) \right\}$$

$$\cong -\left\{ \frac{1}{2N} + 2v[1 - 1 + (\frac{2\pi}{n})^2] \right\}$$

$$= -\left\{ \frac{1}{2N} + 2v(\frac{2\pi}{n})^2 \right\} .$$

Thus these two eigenvalues can be made arbitrarily close to each other by letting n become large. As $n \to \infty$, the spectrum becomes a closed continuous interval, $[-\frac{1}{2N}, -(\frac{1}{2N} + 4v)]$. It is worth noting that the largest eigenvalue is independent of v (mutation rate), (Ohta and Kimura, 1973). We shall next obtain the stationary solution given in (12.16). To do this we need to obtain the inverse matrix of A. We try to expand A^{-1} in terms of $E_i \boxtimes E_j$ matrices where E_i are the eigenvectors of A. i.e.,

$$A^{-1} = \Sigma b_{ij} E_i \boxtimes E_j^*$$

where $\boxtimes$ indicates the direct product of vectors and * indicates the transpose. Then

$$AA^{-1} = A \ \Sigma \ b_{ij} E_i \boxtimes E_j^* = I \ .$$

Note

$$A \ \Sigma \ b_{ij} E_i \boxtimes E_j^* = \Sigma \ b_{ij} \lambda_i E_i \boxtimes E_j^* \ .$$

Thus

$$A \ \Sigma \ b_{ij} E_i \boxtimes E_j = \Sigma \ b_{ij} \lambda_i E_i \boxtimes E_j = I \ .$$

Note

$$E_k^* \cdot [E_i \boxtimes E_j^*] \cdot E_\ell = E_k^* \cdot E_i \boxtimes E_j^* \cdot E_\ell$$

$$= \ \|E_k\|^2 \cdot \|E_\ell\|^2 \quad \text{if } k = i \text{ and } j = \ell,$$

$$= 0 \qquad\qquad \text{if } k \neq i \text{ or } j \neq \ell.$$

Using this property we can determine coefficients b_{ij}:

$$b_{ij} \lambda_i \ \|E_i\|^2 \ \|E_\ell\|^2 = \ \|E_i\|^2 \ \|E_\ell\|^2,$$

$$b_{ij} = \frac{1}{\lambda_i} \ .$$

Therefore

$$A^{-1} = \underset{ij}{\Sigma} \ \frac{1}{\lambda_i} \ E_i \boxtimes E_j^*$$

and

$$-A^{-1} J = \underset{i}{\Sigma} \ \frac{-1}{\lambda_i} \ E_i = \ \frac{1}{n} \ \underset{i}{\Sigma} \ \frac{1}{\frac{1}{2N} + 2v(1 - \cos \frac{2\pi i}{n})} \ (\frac{1}{2N} \ E_i) \ .$$

In other words

$$C(\infty) = \frac{1}{2nN} \ \underset{i}{\Sigma} \ \frac{1}{\frac{1}{2N} + 2v(1 - \cos \frac{2\pi i}{n})} \begin{pmatrix} 1 \\ \cos \frac{2\pi i}{n} \\ \cos \frac{4\pi i}{n} \\ \cdot \\ \cdot \\ \cdot \\ \cos \frac{2(n-1)\pi i}{n} \end{pmatrix}$$

or

$$C_k(\infty) = \frac{1}{2nN} \sum_i \frac{\cos \frac{2k\pi i}{n}}{\frac{1}{2N} + 2v(1 - \cos \frac{2\pi i}{n})}.$$

As $n \to \infty$

$$C_k(\infty) = \frac{1}{4N\pi} \int_0^\pi \frac{\cos k\theta}{\frac{-1}{2N} + 2v(1 - \cos\theta)} d\theta$$

$$= \frac{1}{2\pi} \int_0^\pi \frac{\cos k\theta}{1 + 4Nv(1 - \cos\theta)} d\theta .$$

This integral can be evaluated, and

(12.18)
$$\begin{cases} C_0(\infty) = \dfrac{1}{\sqrt{1+8Nv}} , \\[4mm] C_k(\infty) = \dfrac{1}{\sqrt{1+8Nv}} \left(\dfrac{1 + 4Nv - \sqrt{1+8Nv}}{4Nv} \right)^k . \end{cases}$$

This result was obtained first by Ohta and Kimura (1973).

Now returning to (12.14), we will examine the transient behavior of $C(t)$. Noting $-A^{-1}J = C(\infty)$, we can rewrite (12.14),

(12.19)
$$C(t) = e^{At}\{C(0) - C(\infty)\} + C(\infty).$$

In order to determine this solution explicitly, we need to examine the matrix e^{At} in detail. Noting that matrix $E \equiv [c_{ij}]_{nxn}$ in which $c_{ij} = \cos 2\pi ij/n$ diagonalizes matrix A, we have

$$e^{At} = \frac{1}{n} E \begin{pmatrix} e^{\lambda_0 t} & & & & 0 \\ & e^{\lambda_1 t} & & & \\ & & \cdot & & \\ & & & \cdot & \\ 0 & & & & e^{\lambda_{n-1} t} \end{pmatrix} E .$$

Therefore

$$C(t) = \frac{1}{n} E\Lambda^{(t)} E\{C(0) - C(\infty)\} + C(\infty)$$

where $\Lambda^{(t)}$ is the diagonal matrix on the right side of the above equations. The individual elements of $C(t)$ can be written as

$$C_k(t) = C_k(\infty) + \sum_{i=0}^{n-1} \{C_i(0) - C_i(\infty)\}$$

$$x \sum_{j=0}^{n-1} \cos \frac{2\pi ij}{n} \cos \frac{2\pi kj}{n} e^{-\left\{\frac{1}{2N} + 2v\left(1 - \cos \frac{2\pi j}{n}\right)\right\}}$$

and as $n \to \infty$

(12.20) $$C_k(t) = C_k(\infty) + \sum_{i=-\infty}^{\infty} \{C_i(0) - C_i(\infty)\}$$

$$x \frac{1}{\pi} \int_0^\pi \cos i\xi \, \cos kk\xi \, e^{-\left\{\frac{1}{2N} + 2v(1 - \cos\xi)\right\}} \, d\xi$$

where $C_{-i} = C_i$.

Multiplying both sides of (12.20) by $\cos \theta k$ and adding for all $k = 0, \pm 1, \pm 2, \cdots$, we have

(12.21) $$F_t(\theta) \equiv \sum_{i=-\infty}^{\infty} C_k(t) \cos \theta k$$

$$= \left[1 - e^{-\left\{\frac{1}{2N} + 2v(1-\cos\theta)\right\}t} \right] \frac{1}{1 + 4Nv(1-\cos\theta)}$$

$$+ F_0(\theta) e^{-\left\{\frac{1}{2N} + 2v(1-\cos\theta)\right\}t} .$$

If we differentiate $F_t(\theta)$ twice with respect to θ and let $\theta = 0$, we have the second moment of $C_k(t)$,

$$-F_t''(\theta)\Big|_{\theta=0} = \sum k^2 C_k(t) = 4Nv\left[1 - e^{-\frac{t}{2N}}\right] + 2v[1 + F_0(0)]e^{-\frac{t}{2N}} .$$

As Wehrhahn (1975) has shown, we have

(12.22) $$-F_t''(\theta)\Big|_{\substack{\theta=0 \\ t\to\infty}} = 4Nv = \text{Variance of } C_k \text{ at equilibrium.}$$

Now consider two populations which were split some time in the past, and let $D_k(t)$ be the analogue of $C_k(t)$ for the two populations. Namely, $D_k(t)$ is the probability that two genes, one from each of the populations, are allelic types k-steps apart at t generations after the split. Then we can guess $D_k(t)$ satisfies a differential equation similar to (12.12). Indeed,the only difference for this case of two populations is that there is no possibility that two genes belonging to different populations come from a single gene in the recent past. Therefore if we remove the facter $1/2N$ from (12.12), we have the differential equations for $D_k(t)$:

(12.23)
$$\frac{dD_k(t)}{dt} = v\{D_{k-1}(t) - 2D_k(t) + D_{k+1}(t)\}$$

or

(12.24)
$$\frac{dD(t)}{dt} = BD(t)$$

where $D(t)$ is the column vector consisting of $D_k(t)$ and

(12.25)
$$B = v \begin{pmatrix} -2 & 1 & & & 0 \\ 1 & -2 & 1 & & \\ & \cdot & \cdot & \cdot & \\ & & \cdot & \cdot & 1 \\ 0 & & & 1 & -2 \end{pmatrix} .$$

Symbolically the system of equations (12.24) has the solution given by

(12.26)
$$D(t) = e^{Bt}D(0) .$$

An individual element of $D(t)$ given in (12.26) is

(12.27)
$$D_k(t) = \frac{1}{\pi} \sum_{i=-\infty}^{\infty} D_i(0) \int_0^{\pi} \cos k\xi \cos i\xi \, e^{-2v(1-\cos\xi)t} d\xi .$$

As a special case, if the two populations under consideration have diverged from an equilibrium population and the two were genetically alike at the beginning, we may replace $D_i(0)$ by $C_i(0)$. In that case the formula can be simplified to

(12.28)
$$D_k(t) = \frac{1}{\pi} \sum_{i=-\infty}^{\infty} C_i(\infty) \int_0^{\pi} \cos k\theta \cos i\theta \, e^{-2v(1-\cos\xi)t} d\xi$$
$$= \frac{H_0(1-\lambda^2)}{\pi} \int_0^{\pi} \frac{\cos k\xi}{1 + \lambda^2 - 2\lambda\cos\xi} e^{-2v(1-\cos\xi)t} d\xi$$

where

$$H_0 = 1/\sqrt{1 + 8Nv}$$

and

$$\lambda = \{1 + 4Nv - \sqrt{1 + 8Nv}\}/4Nv .$$

In analogy to (12.21), let

(12.29)
$$G_t(\theta) = \sum_k D_k(t)\cos\theta k .$$

Then, from (12.27) we have

$$G_t(\theta) = G_0(\theta)e^{-2v(1 - \cos\theta)t} .$$

In the case of two population that diverged from an equilibrium population, from (12.28) we have

(12.30) $\quad G_t(\theta) = F_\infty(\theta)e^{-2v(1-\cos\theta)t} = \dfrac{1}{1 + 4Nv(1-\cos\theta)} e^{-2v(1-\cos\theta)t} .$

A special case of (12.30), with $\theta = \pi$, is

$$G_t(\pi) = \frac{1}{1 + 8Nv} e^{-4Nvt} .$$

This is a simple exponential function of t, and therefore taking logarithm we have

(12.31)
$$-\log_e G_t(\pi) = 4vt + \log_e(1 + 8Nv) .$$

which is linear in t. The genetic distance between two populations can be conveniently measured by this quantity, for this increases linearly with time at a rate 4v. Another sepcial case of (12.30) is with $\theta = \pi/2$,

$$G_t(\frac{\pi}{2}) = \frac{1}{1+2Nv} e^{-2vt}$$

and

(12.32)
$$-\log_e G_t(\frac{2\pi}{3}) = 2vt + \log_e(1 + 4Nv) .$$

This special case is due to Wehrhahn (1975). As θ approaches 0, the rate of change in $-\log_e G_t(\theta)$ with time decreases, and for $\theta = 0$

$$G_t(0) = G_0(0).$$

The quantity remains unchanged.

THE INFINITESIMAL MEAN CHANGE ($M_{\delta x}$) AND VARIANCE ($V_{\delta x}$)

This appendix is to provide some detailed calculations of the infinitesimal mean change and variance appearing in the Kolmogorov backward equations (2.5), (2.9), (2.20), (2.22) etc. Derivation of the Kolmogorov equations for discrete and continuous state space are also given. We will use Moran's model for mathematical simplicity.

Mathematically more rigorous treatments for Wright's model are given by Trotter (1958), Watterson (1962), Guess (1973) Ethier (1975) and Sato (1976).

A1.1 Mutation model I

Consider a population consisting of 2N genes of two alleles (A and a) and let the survival probability of each gene be given by the negative exponential law e^{-t}. Assume that every death is immediately followed by a new birth and that every pre-existing gene prior to a death has an equal chance to become a parent. Let the mutation rate from A to a be u and the reverse rate be v.

$$A \underset{v}{\overset{u}{\rightleftarrows}} a$$

In this section we assume that mutation and birth-death are independent. Let x_t be the frequency of A at time t. Consider the time interval (t, $t+\Delta t$). Then the change in x_t due to mutation can be expressed as

(A1.1)
$$x_{t+\Delta t} = x_t + \{v - (u+v)x_t\}\Delta t + o(\Delta t)$$

and

$$1 - x_{t+\Delta t} = 1 - x_t - \{v - (u+v)x_t\}\Delta t + o(\Delta t) .$$

If we calculate the variance without taking the mutational change into account, we have

(A1.2) $\quad \dfrac{1}{\Delta t}\left[2N\Delta t\left\{(\dfrac{1}{2N})^2 x_t(1-x_t) + (\dfrac{1}{2N})^2 x_t(1-x_t)\right\} + o(\Delta t)\right]$

$$\longrightarrow \quad \dfrac{x_t(1-x_t)}{N} \qquad \text{as } \Delta t \to 0 .$$

Now the same calculation using $x_{t+\Delta t}$, instead of x_t, thus taking the mutational change into accout, leads us to

(A1.3) $\quad \dfrac{1}{\Delta t}\left[4N\Delta t(\dfrac{1}{2N})^2 x_{t+\Delta t}(1-x_{t+\Delta t}) + o(\Delta t)\right]$

$$= \dfrac{1}{\Delta t}\left[4N\Delta t\left\{(\dfrac{1}{2N})^2 x_t(1-x_t) + O(\Delta t)\right\} + o(\Delta t)\right]$$

$$\longrightarrow \quad \dfrac{x_t(1-x_t)}{N} \qquad \text{as } \Delta t \to 0 .$$

The infinitesimal variances calculated by the two methods are identical. Therefore, if mutation and birth-death are independent, the infinitesimal variance is not affected and it is given by $x(1-x)/N$.

A1.2 Mutation model II

Assume here that a gene can mutate to the other allele when and only when an individual is born, and that as above the rate from A to a is u and the reverse rate is v. The conditional probabilities of gene frequency change are given in the table below:

Birth \ Death	A, $\quad$ x		a, $\quad$ 1-x		
A, $\quad$ x	0	$-\dfrac{1}{2N}$	0	$\dfrac{1}{2N}$	Amount of change
	$(1-u)x^2$	ux^2	$ux(1-x)$	$(1-u)x(1-x)$	Probability
a, $\quad$ 1-x	0	$-\dfrac{1}{2N}$	0	$\dfrac{1}{2N}$	Amount of change
	$vx(1-x)$	$(1-v)x(1-x)$	$(1-v)(1-x)^2$	$v(1-x)^2$	Probability

And the conditional mean change and its variance are

(A1.4) Mean $(\Delta x) = \frac{1}{2N} [-ux^2 + (1-u)x(1-x) - (1-v)x(1-x) + v(1-x)^2]$

$$= \frac{1}{2N} \{v - (u+v)x\} = \frac{1}{2N} \{-ux + v(1-x)\}$$

and

(A1.5) $E(\Delta x^2) = (\frac{1}{2N})^2 [ux^2 + (1-u)x(1-x) + (1-v)x(1-x) + v(1-x)^2]$

$$= (\frac{1}{2N})^2 [2x(1-x) + (2x-1)\{ux - v(1-x)\}].$$

respectively.

Consider a time interval $(t, t+\Delta t)$. The probabilities of k death-births in a time interval $(t, t+\Delta t)$ are

k	Probability
0	$e^{-2N\Delta t}$
1	$2N\Delta t e^{-2N\Delta t}$
2	$(2N\Delta t)^2 e^{-2N\Delta t}/2!$
.	.
.	.
.	.

Thus the mean $(M_{\Delta x}(\Delta t))$ and the second moment $(V_{\Delta x}(\Delta t))$ of Δx in the time interval $(t, t+\Delta t)$ are

(A1.6) $M_{\Delta x}(\Delta t) = \frac{2N\Delta t}{2N} \{-ux + v(1 - x)\} + o(\Delta t)$

and

(A1.7) $V_{\Delta x}(\Delta t) = 2N\Delta t (\frac{1}{2N})^2 [2x(1-x) + (2x-1)\{ux-v(1-x)\}] + o(\Delta t)$

respectively. Therefore

(A1.8) $$\lim_{\Delta t \to 0} \frac{M_{\Delta x}(\Delta t)}{\Delta t} = -ux + v(1-x) \equiv M_{\delta x},$$

(A1.9) $$\lim_{\Delta t \to 0} \frac{V_{\Delta x}(\Delta t)}{\Delta t} = \frac{x(1-x)}{N} + \frac{1}{2N} \{-ux - v(1-x)\}(2x-1) \equiv V_{\delta x}.$$

The infinitesimal variance given by (A1.9) differs from that of (A1.2) or (A1.3) by a factor of magnitude $(u+v)/2N$.

A1.3 Derivation of Kolmogorov backward equations (KBE)

We shall take an approach of going from a finite matrix system for a discrete model to a continuous differential operator, and see if some of the KBEs (Eq. 2.5; Eq. 2.9; section 2.5; Eq. 2.20; Eq. 2.22 etc.) are valid.

Pure random drift case. Let $p_{ij}^{(t)}$ be the probability that the number of A genes is i at time t, given that it is j at t = 0.

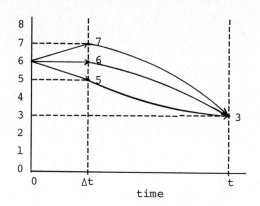

Fig. A1.1 Diagram illustrating paths going
from j = 6 to i = 3 in time interval (0, t).

From this diagram it is easy to see

$$p_{3,6}^{(t)} = (1 - 2N\Delta t)p_{3,6}^{(t-\Delta t)} + 2N\Delta t\left[\frac{6}{2N}(1 - \frac{6}{2N})p_{3,5}^{(t-\Delta t)}\right.$$

$$+ \frac{6}{2N}(1 - \frac{6}{2N})p_{3,7}^{(t-\Delta t)} + \left\{1 - 2(\frac{6}{2N})(1 - \frac{6}{2N})\right\}p_{3,6}^{(t-\Delta t)}\right] + o(\Delta t).$$

In general,

$$p_{ij}^{(t+\Delta t)} = (1 - 2N\Delta t)p_{ij}^{(t)} + 2N\Delta t\left[\frac{j}{2N}(1 - \frac{j}{2N})p_{i,j-1}^{(t)}\right.$$

$$- \left\{1 - 2(\frac{j}{2N})(1 - \frac{j}{2N})\right\}p_{ij}^{(t)} + \frac{j}{2N}(1 - \frac{j}{2N})p_{i,j+1}^{(t)}\right] + o(\Delta t).$$

This can be rearranged to

(A1.10) $$\frac{p_{ij}^{(t+\Delta t)} - p_{ij}^{(t)}}{\Delta t} = \frac{2N\Delta t}{\Delta t}\left\{p_{ij-1}^{(t)} - 2p_{ij}^{(t)} + p_{ij+1}^{(t)}\right\} + \frac{o(\Delta t)}{\Delta t}.$$

Thus

(A1.11)
$$\frac{\partial p_{ij}^{(t)}}{\partial t} = 2N\frac{j}{2N}(1 - \frac{j}{2N}) \left\{ p_{ij-1}^{(t)} - 2p_{ij}^{(t)} + p_{ij+1}^{(t)} \right\}$$

or

(A1.12)
$$\frac{dP^{(t)}}{dt} = 2NP^{(t)} \begin{pmatrix} -2\theta_1 & \theta_1 & & & 0 \\ \theta_2 & -2\theta_2 & \theta_2 & & \\ & & \ddots & \ddots & \ddots \\ 0 & & & \ddots & \ddots & \ddots \end{pmatrix}$$

where $\theta_i = \frac{j}{2N}(1 - \frac{j}{2N})$ and $P^{(t)} = [p_{ij}^{(t)}]$.

If we multiply both side of (A1.11) by 2N and assume that the space between (i, j+1) gets smaller as $N \to \infty$, (A1.11) converges to

$$\frac{\partial p}{\partial \tau} = x(1 - x)\frac{\partial^2 p}{\partial x^2}$$

where $\tau = 2Nt$ or

(A1.13)
$$\frac{\partial u}{\partial t} = \frac{x(1 - x)}{2N} \frac{\partial^2 u}{\partial x^2}$$

where $u = u(t, x) = \int p(t, x, y)f(y)dy$. This is KBE (2.9).

Mutation model I

From the diagram in Fig. A1.1, we can see

$$p_{3,6}^{(t)} = (1 - 2N\Delta t - iu\Delta t - (2N - j)v\Delta t)p_{3,6}^{(t-\Delta t)}$$

$$+ 2N\Delta t \begin{bmatrix} \text{Death-birth term,} \\ \text{the same as in the} \\ \text{no mutation case} \end{bmatrix}$$

$$+ 6u\Delta t p_{3,5}^{(t-\Delta t)} + (2N - 6)v\Delta t p_{3,7}^{(t-\Delta t)} + o(\Delta t) .$$

In general,

(A1.14)
$$p_{ij}^{(t+\Delta t)} = \{1 - 2N\Delta t - iu\Delta t - (2N - j)v\Delta t\}p_{ij}^{(t)}$$

$$+ 2N\Delta t \quad [\text{term due to death-birth}]$$

$$+ iu\Delta t p_{ij-1}^{(t)} + (2N - i)v\Delta t p_{ij+1}^{(t)} + o(\Delta t) .$$

From this,

(A1.15) $\dfrac{\partial P_{ij}^{(t)}}{\partial t} = 2N\dfrac{j}{2N}(1 - \dfrac{j}{2N})\left\{P_{ij-1}^{(t)} - 2P_{ij}^{(t)} + P_{ij+1}^{(t)}\right\} + \dfrac{2Nju}{2N}P_{ij-1}^{(t)}$

$$- \left\{\dfrac{2Nju}{2N} + \dfrac{2N(2N-j)v}{2N}\right\}P_{ij}^{(t)} + \dfrac{2N(2N-j)v}{2N}P_{ij+1}^{(t)}$$

or

(A1.16)
$$\frac{dP^{(t)}}{dt} = 2NP^{(t)}\left[\begin{pmatrix} -2\theta_1 & +\theta_1 & & & \mathbf{0} \\ \theta_2 & -2\theta_2 & \theta_2 & & \\ & & \ddots & \ddots & \ddots \\ \mathbf{0} & & & & \end{pmatrix}\right.$$

$$\left.+ \begin{pmatrix} -(\alpha_1+\beta_1), & \beta_1 & & & \mathbf{0} \\ \alpha_2, & -(\alpha_2+\beta_2), & \beta_2 & & \\ & \ddots & \ddots & \ddots & \\ \mathbf{0} & & \ddots & & \end{pmatrix}\right].$$

Note that

$$x(1 - x)\lim_{2N\to\infty} \frac{1}{(\frac{1}{2N})^2}\left\{P_{ij-1}^{(t)} - 2P_{ij}^{(t)} + P_{ij+1}^{(t)}\right\}$$

$$\longrightarrow x(1 - x)\frac{\partial^2}{\partial x^2}p$$

and

$$xu\lim_{2N\to\infty} \frac{P_{ij-1}^{(t)} - P_{ij}^{(t)}}{(\frac{1}{2N})} + (1 - x)v\lim_{2N\to\infty} \frac{P_{ij}^{(t)} + P_{ij+1}^{(t)}}{(\frac{1}{2N})}$$

$$= -xu\frac{\partial}{\partial x}p + (1 - x)v\frac{\partial}{\partial x}p .$$

Hence multiplying both sides of (A1.15) by 2N and letting N become large while keeping Nu and Nv constant, we have

(A1.17) $\dfrac{\partial P}{\partial \tau} = x(1 - x)\dfrac{\partial^2 P}{\partial x^2} + \{-Ux + V(1 - x)\}\dfrac{\partial P}{\partial x}$

where $P = P(\tau, x, y)$ with $\tau = 2Nt$, $U = 2Nu$ and $V = 2Nv$.

Selection model (genic selection)

$$
\begin{array}{ccc}
\text{Gene} & A & a \\
\text{Fitness} & 1 + s & 1
\end{array}
$$

The conditional probabilities of gene frequency change are given in the table below:

Birth / Death	A $\frac{j}{2N}$	a $1 - \frac{j}{2N}$	
$A,\ \dfrac{j+sj}{2N+sj}$	0	$+1/2N$	Amount of change
	$\dfrac{j}{2N}\left(\dfrac{j+sj}{2N+sj}\right)$	$\left(1-\dfrac{j}{2N}\right)\left(\dfrac{j+sj}{2N+sj}\right)$	Probability
$a,\ \dfrac{2N-j}{2N+sj}$	$-1/2N$	0	Amount of change
	$\dfrac{j}{2N}\left(\dfrac{2N-j}{2N+sj}\right)$	$\left(1-\dfrac{j}{2N}\right)\left(\dfrac{2N-j}{2N+sj}\right)$	Probability

From this scheme of death-birth and selection,

$$
(A1.18) \qquad p_{ij}^{(t+\Delta t)} = (1 - 2N\Delta t)p_{ij}^{(t)} + 2N\Delta t \left[\frac{-j}{2N}\left(\frac{2N-j}{2N+sj}\right)p_{ij-1}^{(t)} \right.
$$

$$
+ \left(1 - \frac{j}{2N}\right)\left(\frac{j+sj}{2N+sj}\right)p_{ij+1}^{(t)}
$$

$$
\left. + \left\{\frac{j}{2N}\left(\frac{j+sj}{2N+sj}\right) + \left(1 - \frac{j}{2N}\right)\left(\frac{2N-j}{2N+sj}\right)\right\}p_{ij}^{(t)} \right] + o(\Delta t).
$$

Note

$$
\left(\frac{j}{2N}\right)\left(\frac{2N-j}{2N+sj}\right) = \left(\frac{j}{2N}\right)\left(1 - \frac{j}{2N}\right) - s\left(\frac{j}{2N}\right)^2\left(1 - \frac{j}{2N}\right) + O(s^2),
$$

$$
\left(1-\frac{j}{2N}\right)\left(\frac{j+sj}{2N+sj}\right) = \left(\frac{j}{2N}\right)\left(1 - \frac{j}{2N}\right) + s\left(\frac{j}{2N}\right)\left(1 - \frac{j}{2N}\right)^2 + O(s^2),
$$

$$
\left(\frac{j}{2N}\right)\left(\frac{j+sj}{2N+sj}\right) = \left(\frac{j}{2N}\right)^2 + s\left(\frac{j}{2N}\right)^2\left(1 - \frac{j}{2N}\right) + O(s^2),
$$

$$
\left(1-\frac{j}{2N}\right)\left(\frac{2N-j}{2N+sj}\right) = \left(1 - \frac{j}{2N}\right)^2 - s\left(\frac{j}{2N}\right)\left(1 - \frac{j}{2N}\right)^2 + O(s^2),
$$

and

$$
\left(\frac{j}{2N}\right)^2 + \left(1 - \frac{j}{2N}\right)^2 - 1 = -2\left(\frac{j}{2N}\right)\left(1 - \frac{j}{2N}\right).
$$

Using these identities, equation (A1.18) can be rearranged to

$$\frac{p_{ij}^{(t+\Delta t)} - p_{ij}^{(t)}}{\Delta t} = 2N(\tfrac{j}{2N})(1-\tfrac{j}{2N})\left\{p_{ij-1}^{(t)} - 2p_{ij}^{(t)} + p_{ij+1}^{(t)}\right\}$$

$$- 2Ns(\tfrac{j}{2N})^2(1-\tfrac{j}{2N})p_{ij-1}^{(t)} + 2Ns(\tfrac{j}{2N})(1-\tfrac{j}{2N})^2 p_{ij+1}^{(t)}$$

$$+ 2Ns(\tfrac{j}{2N})^2(1-\tfrac{j}{2N})p_{ij}^{(t)} - 2Ns(\tfrac{j}{2N})(1-\tfrac{j}{2N})^2 p_{ij}^{(t)}$$

$$+ 2N\, O\,(s^2) + o(\Delta t)/\Delta t .$$

Therefore as $\Delta t \to 0$,

$$(A1.19) \qquad \frac{\partial p_{ij}^{(t)}}{\partial t} = 2N(\tfrac{j}{2N})(1 - \tfrac{j}{2N})\left\{p_{ij-1}^{(t)} - 2p_{ij}^{(t)} + p_{ij+1}^{(t)}\right\}$$

$$+ 2Ns(\tfrac{j}{2N})^2(1 - \tfrac{j}{2N})\left\{p_{ij}^{(t)} - p_{ij-1}^{(t)}\right\}$$

$$+ 2Ns(\tfrac{j}{2N})(1 - \tfrac{j}{2N})^2\left\{p_{ij+1}^{(t)} - p_{ij}^{(t)}\right\} + 2N\,O(s^2) .$$

Divide both sides of Eq. (A1.19) by $1/2N$ and let $N \to \infty$ keeping $S = Ns$ constant. (Note $2N\,O(s^2) \to 0$ as $N \to \infty$) Hence we have,

$$\frac{\partial u}{\partial \tau} = x(1 - x)\frac{\partial^2 P}{\partial x^2} + 2Sx\frac{\partial P}{\partial x} + 2S(1 - x)\frac{\partial P}{\partial x}$$

$$= x(1 - x)\frac{\partial^2 P}{\partial x^2} + 2Sx(1 - x)\frac{\partial P}{\partial x} .$$

where $P = P(\tau, x, y)$ with $\tau = 2Nt$.
If we let $u(\tau, x) = \int P(\tau, x, y)f(y)dy$, we have

$$(A1.20) \qquad \frac{\partial u(\tau, x)}{\partial t} = x(1 - x)\frac{\partial^2 u(\tau, x)}{\partial x^2} + Sx(1 - x)\frac{\partial u(\tau, x)}{\partial x} .$$

And if we let $t = \tau/2N$,

$$\frac{\partial u(t, x)}{\partial t} = \frac{x(1 - x)}{2N}\frac{\partial^2 u(t, x)}{\partial x^2} + sx(1 - x)\frac{\partial u(t, x)}{\partial x} .$$

Mutation model II

The diagram below illustrates the transition from j to i in time $t+\Delta t$.

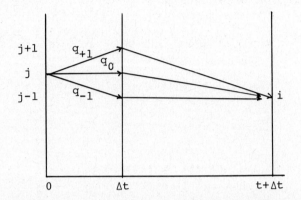

Fig. A1.2 Diagram illustrating possibilities
going from j to i, via three intermediate
steps, $j+1(q_{+1})$, $j(q_0)$ and $j-1(q_{-1})$.

These probabilities are

$$q_0 = (1 - 2N\Delta t) + 2N\Delta t[(1-u)x^2 + ux(1-x)$$
$$+ vx(1-x) + (1-v)(1-x)^2] + o(\Delta t),$$

$$q_{-1} = 2N\Delta t[ux^2 + (1-v)x(1-x)] + o(\Delta t),$$

$$q_{+1} = 2N\Delta t[(1-u)x(1-x) + v(1-x)^2] + o(\Delta t).$$

Let $p_{ij}^{(t)}$ be the probability that the number of A genes is i at time t,
given that it is j at t = 0. Let us calculate the recurrence relationship for $p_{ij}^{(t)}$. For convenience denote j/2N by x. Then

(A1.21) $$p_{ij}^{(t+\Delta t)} = (1-2N\Delta t)p_{ij}^{(t)} + 2N\Delta t\{(1-u)x^2 + ux(1-x)$$
$$+ vx(1-x) + (1-v)(1-x)^2\}p_{ij}^{(t)}$$
$$+ 2N\Delta t\{ux^2 + (1-v)x(1-x)\}p_{i,j-1}^{(t)}$$
$$+ 2N\Delta t\{(1-u)x(1-x) + v(1-x)^2\}p_{i,j+1}^{(t)} + o(\Delta t).$$

The first term on the right side of Eq. (A1.21) accounts for the possibility that no death-birth occurs. The second term is that one
death-birth occurs, but the number of A genes remains unchanged. The

third term is that one death-birth occurs and it is decreased by one. The fourth term is the possibility of increasing by one. The last term takes care of all higher order events.

Eq. (A1.21) can be rearranged to

(A1.22)
$$p_{ij}^{(t+\Delta t)} - p_{ij}^{(t)} = 2N\Delta t \left[\{x(1-x) - vx(1-x) + ux^2\}p_{i,j-1}^{(t)} \right.$$
$$+ \{-2x(1-x) + vx(1-x) - ux^2 + ux(1-x) - v(1-x)^2\}p_{ij}^{(t)}$$
$$+ \left. \{x(1-x) - ux(1-x) + v(1-x)^2\}p_{i,j+1}^{(t)} \right] + o(\Delta t).$$

Dividing both sides of Eq. (A1.22) by Δt and letting $\Delta t \to 0$, we have

(A1.23)
$$\frac{dp_{ij}^{(t)}}{dt} = 2N\left[\{x(1-x) - vx(1-x) + ux^2\}p_{ij-1}^{(t)} \right.$$
$$+ \{-2x(1-x) + vx(1-x) - ux^2 + ux(1-x) - v(1-x)^2\}p_{ij}^{(t)}$$
$$+ \left. \{x(1-x) - ux(1-x) + v(1-x)^2\}p_{ij+1}^{(t)} \right]$$

where $x = j/2N$.

This equation can be written as

(A1.24)
$$\frac{dQ^{(t)}}{dt} = 2NQ^{(t)} \left[\begin{pmatrix} -2\theta_0 & \theta_0 & & \text{O} \\ \theta_1 & \theta_1 & \theta_1 & \\ & & \cdot & \cdot & \cdot \\ \text{O} & & \cdot & \cdot & \cdot \end{pmatrix} \right.$$

$$+ \left. \begin{pmatrix} -\alpha_0 - \beta_0, & \beta_0 & & \text{O} \\ \alpha_1, & -\alpha_1 - \beta_1, & \beta_1 & \\ & & \cdot & \cdot & \cdot \\ \text{O} & & \cdot & \cdot & \cdot \end{pmatrix} \right]$$

where $\theta_i = x_i(1-x_i)$, $\alpha_i = -vx_i(1-x_i) + ux_i^2$ and $\beta_i = -ux_i(1-x_i) + v(1-x_i)^2$. For convenience, rewrite Eq. (A1.24) as

$$\text{(A1.25)} \qquad \frac{dP^{(t)}}{dt} = 2NP^{(t)}[A + B]$$

where A and B are respectively the first and second matrix appearing on the right side of Eq. (A1.24).

For any given vector $w = (w_0, w_1, w_2, \cdots)$ define $u^{(t)} = wP^{(t)}$. Then from (A1.25), we have

$$\text{(A1.26)} \qquad \frac{du^{(t)}}{dt} = 2Nu^{(t)}[A + B] .$$

We multiply both sides of Eq. (A1.26) by 2N and write the resulting equation as

$$\text{(A1.27)} \qquad 2N \frac{du^{(t)}}{dt} = u^{(t)} \frac{1}{(\frac{1}{2N})^2} A + \frac{2N}{(\frac{1}{2N})} B .$$

If we view $u^{(t)}$ as a function defined on [0, 1], and $\frac{1}{(\frac{1}{2N})^2} A \quad \frac{1}{(\frac{1}{2N})} B$ as operators on such functions, and let $2N \to \infty$ while keeping 2Nu and 2Nv constants, we have

$$\text{(A1.28)} \qquad \frac{1}{(\frac{1}{2N})^2} A \longrightarrow x(1 - x)\frac{\partial^2}{\partial x^2}$$

and

$$\text{(A1.29)} \qquad \frac{1}{(\frac{1}{2N})} B \longrightarrow \{2Nvx(1-x) - 2Nux^2 - 2Nux(1-x) + 2Nv(1-x)^2\}\frac{\partial}{\partial x}$$

$$= \{-2Nux + 2Nv(1-x)\}\frac{\partial}{\partial x} .$$

Therefore at such a limit, Eq. (A1.27) formally converges to

$$\text{(A1.30)} \qquad \frac{\partial u(\tau, x)}{\partial \tau} = x(1-x)\frac{\partial^2}{\partial x^2} u(\tau, x) + \{-Ux + V(1-x)\}\frac{\partial}{\partial x} u(\tau, x)$$

where $U = 2Nu$ and $V = 2Nv$ and $\tau = 2Nt$. Hence this equation is the same as (A1.17) of Model I if we substitute u(t, x) for P(t, x, y) in (A1.17) This may appear to be inconsistent with the infinitesimal variances for the two mutational models given by (A1.3) and (A1.9). The two cases differ by a factor proportional to (u+v)/2N. But since we assume 2Nu = U and 2Nv = V stay constant, 2N{(u+v)/2N} diminishes as N gets large and the limiting equations are the same for the two models. Reference: Karlin and McGregor (1964).

APPENDEX II

A SUPPLEMENTARY NOTE ON THE EXISTENCE AND UNIQUENESS OF
THE SOLUTION FOR THE RECURRENCE EQUATION (9.7)

We shall first build up some powerful mathematical machinery, and then prove that iteration (recurrence) equation (9.7) has indeed one and only one solution which is stationary and iterations with any initial vector converges to the unique stationary solution.

Definition: A set X is called a metric space, if there is defined a real valued function $d(x, y)$ for every pair $x, y \in X$, such that
1) $d(x, y) \geq 0$ and $d(x, y) = 0$ if and only if $x = y$;
2) $d(x, y) = d(y, x)$;
3) $d(x, z) \leq d(x, y) + d(y, z)$, triangular inequality.
   $d(,)$ is called the metric of X.

Definition: Let A be a mapping from a metric space X into X. Mapping A is called a contraction mapping, if for every pair $x, y \in X$,

$$d(Ax, Ay) \leq a\, d(x, y)$$

for a fixed constant $a < 1$.

Theorem (The principle of contraction mapping): In a complete* metric space, a contraction mapping has one and only one fixed point, that is

$$Ax = x \ .$$

---

* Complete metric space is a space where every fundamental sequence $\{x_m\}$ converges to a limit which belongs to the space. A fundamental sequence is a sequence which has the following property: For every $\varepsilon > 0$ there is a constant N such that $d(x_n, x_m) < \varepsilon$ for every $n, m > N$.

Furthermore, if we let for any x_0, $x_1 = Ax_0$, $x_2 = Ax_1$, $x_3 = Ax_2$ and so on, then x_n converges to the fixed point, that is

$$\lim_{n \to \infty} Ax_n = \lim_{n \to \infty} x_{n+1} = x = Ax .$$

Therefore, an iteration starting any x_0 converges to the fixed point.

Problem 1. Prove that the real line R^1 with metric $d(x,y) = |x-y|$ is a complete metric space.

Problem 2. Prove that n-dimensional space R^n with metric

$$d(x, y)_p = \left[\sum_i |x_i - y_i|^p \right]^{\frac{1}{p}} \text{ for } p \geq 1 \text{ is a complete metric space.}$$

Here $x = (x_1, x_2, \cdots, x_n)$ and $y = (y_1, y_2, \cdots, y_n)$. In particular

$$d(x, y)_1 = \sum_{i=1}^{n} |x_i - y_i| ,$$

$$d(x, y)_2 = \left[\sum_{i=1}^{n} |x_i - y_i|^2 \right]^{\frac{1}{2}} ,$$

$$d(x, y)_\infty = \max_i |x_i - y_i| .$$

Proof of the theorem. Let x_0 be an arbitrary point in X, and $x_1 = Ax_0$, $x_2 = Ax_1 = A^2 x_0$, $\cdots$, $x_n = Ax_{n-1} = \cdots = A^n x_0$. For $m \geq n$,

$$d(x_n, x_m) = d(A^n x_0, A^m x_0) \leq a^n d(x_0, x_{m-n})$$

$$\leq a^n [d(x_0, x_1) + d(x_1, x_2) + \cdots + d(x_{m-n-1}, x_{m-n})]$$

$$\leq a^n d(x_0, x_1)[1 + a + a^2 + \cdots + a^{m-n-1}]$$

$$\leq a^n d(x_0, x_1) \frac{1}{1-a} .$$

Since $a < 1$, the righthand side goes to 0. Therefore, $\{x_n = A^n x_0\}$ is a fundamental sequence. By the assumption of the completeness of space X, $\lim_{n \to \infty} x_n = x$ exists. Hence

$$Ax = A \lim_{n \to \infty} x_n = \lim_{n \to \infty} Ax_n$$

$$= \lim_{n \to \infty} x_n = x .$$

Thus, $Ax = x$, (existence of a fixed point).

If there are two fixed points

$$Ax = x \quad \text{and} \quad Ay = y, \quad \text{then}$$

$$d(x, y) = d(Ax, Ay) \leq ad(x, y).$$

Hence $d(x, y) = 0$ and $x = y$, (uniqueness).

Applying the contraction principle, we can now prove the existence and uniqueness of iteration equation (9.7).

Equation (9.7) is certainly a mapping from R^n into R^n. The equation is

(A2.1)
$$F_{t+1} = (1 - u)^2 M[F_t + G_t]$$

where

$$F_t = (f_{t,o}, \; f_{t,1}, \; \cdots, \; f_{t,n-1})^*, \qquad (* = \text{transpose}),$$

$$G_t = (\frac{1}{2N} - \frac{f_{t,0}}{2N}, \; 0, \; 0, \; \cdots, \; 0)^*,$$

$$M = (\frac{m}{2})^2 R^{-2} + m(1 - m)R^{-1} + \{(1 - m)^2 + \frac{m^2}{2}\}I$$
$$+ m(1 - m)R + (\frac{m}{2})^2 R^2. \qquad (\text{see } 9.3, \text{ p.132})$$

Now rewrite equation (A2.1) as

(A2.2)
$$F_{t+1} = (1 - u)^2 [M - \underset{\sim}{N}]F_t + q = [a_{ij}]F_t + q$$

where

$$\underset{\sim}{N} = \begin{bmatrix} \frac{1}{2N} & 0 & \cdots & 0 \\ & & & \\ 0 & & O & \\ & & & \end{bmatrix},$$

$$q = (\frac{1}{2N}, \; 0, \; 0, \; \cdots \; 0)^*.$$

Note that

(A2.3)
$$[a_{ij}], \; 'a_{11} = (1 - u)^2 (1 - \frac{1}{2N}),$$

$$a_{ij} = (1 - u)^2 \text{ for } i \neq 1 \text{ or } j \neq 1.$$

Formula (A2.2) certainly defines a mapping from R^n into R^n. Is this a contraction mapping?

For any x, $y \in R^n$ let

(A2.4)
$$x' = [a_{ij}]x + q,$$
$$y' = [a_{ij}]y + q.$$

Let us use the maximum metric $d(x, y)_\infty = \max_i |x_i - y_i|$ to examine whether (A2.4) is indeed a contraction mapping. Then by the definition,

$$d(x', y')_\infty = \max_i |x_i' - y_i'| \le \max_i |\sum_j a_{ij}(x_j - y_j)|$$

$$\le \max_i \sum_j |a_{ij}||x_j - y_j| \le \max_i \sum_j |a_{ij}| \max_j |x_j - y_j|$$

$$= \max_i \sum_j |a_{ij}| d(x, y).$$

Note that $\max_i \sum_j |a_{ij}| = (1 - u)^2 < 1$, (see (A2.3)). Hence

$$d(x', y') \le (1 - u)^2 d(x, y)$$

which proves mapping (A2.4) is indeed a contraction mapping.

Now the theorem of contraction principle assures that recurrence (or iteration) given by (A2.2) converges to the unique, stationary solution for an arbitrary initial vector.

Problem 3: Carry out an analogous proof using another metric such as $d(x, y)_1 = \sum_{i=1}^n |x_i - y_i|$ or $d(x, y)_2 = \left[\sum_{i=1}^n |x_i - y_i|^2 \right]^{\frac{1}{2}}$.

Problem 4: Provide a proof of the existence and uniqueness for the iteration scheme given by Equation (9.24) for a continuous space model. Here the metric space is a collection of functions and we may choose any reasonable function space. As a norm there are several choices, such as

$$d(f, g) = \max_x |f(x) - g(x)|,$$

$$d(f, g) = \int |f(x) - g(x)| dx,$$

or more generally

$$d(f, g) = \left[\int |f(x) - g(z)|^p dx \right]^{\frac{1}{p}} \quad \text{for } p > 1.$$

Nagylaki (1977) has shown that asymptotically the recurrence equation (9.7) converges to its equilibrium at a rate of $\exp\left\{-(2u + \frac{1}{2N_T})t\right\}$ in which u, N_T and t are the mutation rate, total population size and time in generations. The asymptotic rate is independent of the dimensionality of the space.

APPENDIX III

DISTRIBUTION OF STOCHASTIC INTEGRALS

In chapter 4, we dealt with the expectations and the higher moments of stochastic integrals of a given quantity for a given process. But the distributions of the integrals were not obtained, except the special case of sojourn time. In theory the distribution can be determined from the moments, but in practice it is often quite difficult to compute higher moments. Therefore it is of practical importance to develop a feasible method that overcomes this difficulty. This appendix is to provide theoretical and numerical methods that enable us to obtain the distribution of such a stochastic integral. The integral of interest to us is given by (4.71):

$$(A3.1) \qquad \int_0^{T_\omega} f(x_\xi, \ \omega) d\xi$$

where $f(x)$ is an arbitrary function, ω is a sample path and T_ω is the time at which path ω is disregarded from our attention. (Notations in (4.71) and (A3.1) are different; τ_ω instead of T_ω is used in (4.71).) For a given process for which sample paths are designated by $\{\omega\}$, let

$$(A3.2) \qquad \tilde{u}(\tau, \ x) \ = \ \text{prob.} \left\{ \int_0^{\tau_\omega} f(x_\xi, \ \omega) d\xi \geq \tau \right\}.$$

In other words, $\tilde{u}(\tau, \ x)$ is the probability that the value of the quantity given by the above formula (A3.1) is greater than τ, given that the initial gene frequency is x.

Then $\tilde{u}(\tau, \ x)$ can be regarded as a process obtained by a random time change

(A3.3)
$$t = \int_0^{\tau_\omega} f(x_\xi, \omega) d\xi$$

and $\tilde{u}(\tau, x)$ satisfies the KBE

(A3.4)
$$\frac{\partial \tilde{u}(\tau, x)}{\partial \tau} = \frac{1}{f(x)} A\tilde{u}(\tau, x) \equiv \tilde{A}\tilde{u}(\tau, x),$$

$$\tilde{u}(0, x) = 1,$$

where A is the operator for the (original) process. See (5.33) and (5.34). The appropriate solution of (A3.4) is the distribution density of the integral given by (A3.1) as a function of the initial gene frequency x.

For instance, if the quantity to be measured is the heterozygosity, $f(x) = 2x(1-x)$ and therefore the distribution $\tilde{u}(\tau, x)$ satisfies

(A3.5)
$$\frac{\partial \tilde{u}(\tau, x)}{\partial \tau} = \frac{1}{2x(1-x)} A\tilde{u}(\tau, x)$$

$$= \frac{1}{2x(1-x)} \left\{ \frac{V_{\delta x}}{2} \frac{\partial^2}{\partial x^2} + M_{\delta x} \frac{\partial}{\partial x} \right\} \tilde{u}(\tau, x),$$

with the initial condition $\tilde{u}(0, x) = 1$. In particular, for the genic selection model with exit boundaries at both ends of the state space, the KBE is

(A3.6)
$$\frac{\partial \tilde{u}(\tau, x)}{\partial \tau} = \frac{1}{2x(1-x)} \left\{ \frac{x(1-x)}{4N} \frac{\partial^2}{\partial x^2} + sx(1-x)\frac{\partial}{\partial x} \right\} \tilde{u}(\tau, x)$$

$$= \left\{ \frac{1}{8N} \frac{\partial^2}{\partial x^2} + \frac{s}{2} \frac{\partial}{\partial x} \right\} \tilde{u}(\tau, x),$$

and

(A3.7)
$$\tilde{u}(0, x) = 1 \qquad \text{for } 0 < x < 1, \text{ and}$$

$$\tilde{u}(\tau, 0) = \tilde{u}(\tau, 1) = 0.$$

The solution of (A3.6) which satisfies the initial and the boundary conditions (A3.7) is

(A3.8) $$\tilde{u}(\tau, x) = 2e^{-2Sx} \sum_{n=1}^{\infty} \frac{[n\pi\{1-e^{2S}\cos n\pi\}] \sin n\pi x}{(2S)^2 + (n\pi)^2} \exp\left\{ -\frac{n^2\pi^2+4S^2}{8N} \tau \right\}$$

where $S = Ns$, and in particular if $S = 0$,

(A3.9) $$\tilde{u}(\tau, x) = \frac{4}{\pi} \sum_{n=1}^{\infty} \frac{\sin(2n-1)\pi x}{(2n-1)} \exp\left\{-\frac{(2n-1)^2\pi^2}{8N}\tau\right\}.$$

The value of $\tilde{u}(\tau, x)$ is the probability that the sum of heterozygosity along a sample path is greater than τ. The average heterozygosity should be equal to

$$\int_0^{\infty} \tilde{u}(\tau, x)d\tau .$$

Indeed, if $s = 0$,

$$\int_0^{\infty} \tilde{u}(\tau, x)d\tau = \frac{4}{\pi} \sum_{n=1}^{\infty} \frac{\sin(2n-1)\pi x}{(2n-1)} \exp\left\{-\frac{(2n-1)^2\pi^2}{8N}\tau\right\}$$

$$= \frac{32N}{\pi^3} \sum_{n=1}^{\infty} \frac{\sin(2n-1)\pi x}{(2n-1)^3}$$

$$= \frac{32N}{\pi^3} \frac{\pi^3 x(1-x)}{8} = 4Nx(1-x) .$$

This is in agreement with previous results discussed in chapter 4. Fig. A3.1 shows the distribution $\tilde{u}(\tau, x)$ for several cases of the genic selection model.

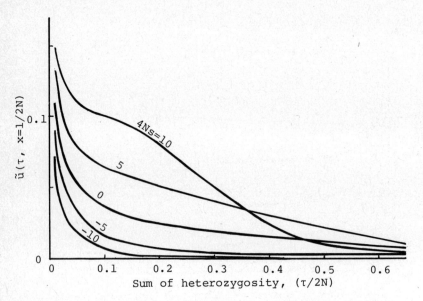

Fig. A3.1 Distribution of heterozygosity
for x = 1/2N. Genic selection.

If the quantity to be measured is the total number of mutant genes and if we consider the genic selection case,

$$f(x) = Nx$$

and $\tilde{u}(\tau, x)$ satisfies

(A3.10) $$\frac{\partial \tilde{u}(\tau, x)}{\partial \tau} = \frac{1}{Nx}\left\{\frac{x(1-x)}{4N}\frac{\partial^2 \tilde{u}(\tau, x)}{\partial x^2} + sx(1-x)\frac{\partial \tilde{u}(\tau, x)}{\partial x}\right\}$$

$$= \left\{\frac{(1-x)}{4N^2}\frac{\partial^2}{\partial x^2} + s(1-x)\frac{\partial}{\partial x}\right\}\tilde{u}(\tau, x)$$

with $\tilde{u}(0, x) = 1$ for $0 < x < 1$ and $\tilde{u}(\tau, 0) = \tilde{u}(\tau, 1) = 0$. The solution of (A3.10) which satisfies the above boundary and initial conditions gives the probability that the total number of mutant genes is greater than τ, given that a path starts from x. If we are to measure the total number of affected individuals,

$$f(x) = N\{x^2 + 2x(1 - x)\}$$

and

$$\frac{\partial \tilde{u}(\tau, x)}{\partial \tau} = \frac{1}{N[x^2+2x(1-x)]}\left\{\frac{V_{\delta x}}{2}\frac{\partial^2}{\partial x^2} + M_{\delta x}\frac{\partial}{\partial x}\right\}\tilde{u}(\tau, x) .$$

Fig. A3.2 and A3.3 show some examples of the affected individuals.

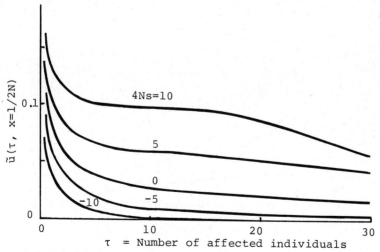

Fig. A3.2 Distribution of number of affected individuals. Genic selection (h=0.5).

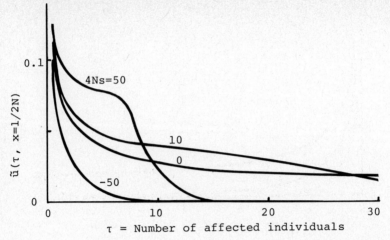

Fig. A3.3 Distribution of number of affected
individuals. Completely recessive mutant
(h=0).

Let us next consider the infinite allele model and calculate the distribution of the heterozygosity. The KBE for this case is

$$(A3.11) \quad \frac{\partial \tilde{u}(\tau, x)}{\partial \tau} = \frac{1}{2x(1-x)} \left\{ \frac{x(1-x)}{4N} \frac{\partial^2}{\partial x^2} - ux\frac{\partial}{\partial x} \right\} \tilde{u}(\tau, x)$$

$$= \left\{ \frac{1}{8N} \frac{\partial^2}{\partial x^2} - \frac{u}{2(1-x)} \frac{\partial}{\partial x} \right\} \tilde{u}(\tau, x),$$

and the boundary and the initial conditions are

$$\tilde{u}(0, x) = 1,$$
$$(A3.12) \qquad \tilde{u}(\tau, 0) = 0 \quad \text{and}$$
$$\frac{\partial \tilde{u}(\tau, x)}{\partial x} \bigg|_{x=1} = 0.$$

Fig. A3.4 shows some examples of the heterozygosity distributions calculated from (A3.11) and (A3.12).

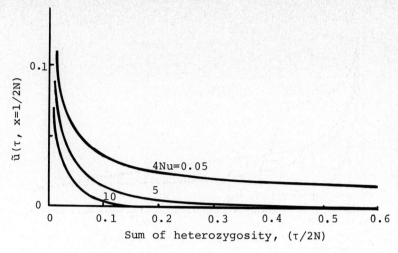

Fig. A3.4 Distribution of heterozygosity
for x = 1/2N. The infinite allele model.

REFERENCES

Abramowitz, M. and I. A. Stegun. (1964). Handbook of mathematical
   functions with formulas, graphs and mathematical tables.
   Washington D. C.: U. S. Department of Commerce.

Ahlfors, L. V. (1953). Complex analysis. An introduction to the
   theory of analytic functions of one complex variable. New York:
   McGraw-Hill.

Coddington, E. A. and N. Levinson. (1955). Theory of ordinary
   differential equations. New York: McGraw-Hill.

Courant, R. and D. Hilbert. (1962). Methods of mathematical physics.
   Vol. I. New York: Interscience.

Crow, J. F. and M. Kimura. (1970). An introduction to population
   genetics theory. New York: Harper and Row.

Crow, J. F. and T. Maruyama. (1971). The number of neutral alleles
   maintained in a finite, geographically structured population.
   Theoret. Popul. Biol. 2: 437-453.

Darling, D. A. and A. I. Siegert. (1953). The first passage problem
   for a continuous Markov process. Ann. Math. Statist. 24: 624-639.

Dynkin, E. B. (1965). Markov processes. Berlin-Gottingen-Heidelberg:
   Springer-Verlag.

Erdélyi, A. (1953). Higher transcendental functions. Vol. II.
   New York: McGraw-Hill.

Ethier, S. (1975). An error estimate for the diffusion approximation
   in population genetics. Ph. D. Thesis, Madison, Wisconsin:
   University of Wisconsin Press.

Ewens, W. J. (1963). The diffusion equation and a pseudo-distribution
   in genetics. J. Roy. Statist. Soc. B. 25: 405-412.

Ewens, W. J. (1969). Population genetics. London: Methuen.

Ewens, W. J. (1972). The sampling theory of selectively neutral
   alleles. Theoret. Popul. Biol. 3: 87-112.

Feller, W. (1951). Diffusion processes in genetics. Proc. 2nd
   Berkeley Symp. Math. Statist. Probab., pp. 227-246. Berkeley:
   University of California Press.

Feller, W. (1952). The parabolic differential equations and the
   associated semi-groups of transformations. Ann. Math. 55:
   468-519.

Feller, W. (1954). Diffusion processes in one dimension. Trans. Am. Math. Soc. 77: 1-31.

Feller, W. (1965). An introduction to probability theory and its applications. Vol. 2. New York: John Wiley.

Fisher, R. A. (1930). The genetical theory of natural selection. Oxford: Calrendon Press. 2nd revised edition, New York: Dover Publication, (1958).

Goursat, E. (1915). Cours d'analyse mathématique, Tome II. Gauthier-Villars, Paris. A course in mathematical analysis, translated by E. R. Hendrick and O. Dunkel, Vol. II. part 2, New York: Dover publication.

Guess, H. A. (1973). On the weak convergence of Wright-Fisher models. Stochastic Processes and Applications 1: 287-306.

Haldane, J. B. S. (1927). A mathematical theory of natural and artificail selection. V. Selection and mutation. Proc. Cambridge Phil. Soc. 23: 838-844.

Imaizumi, Y., N. E. Morton and D. E. Harris. (1970). Isolation by distance in artificial populations. Genetics 66: 569-582.

Ince, E. L. (1927). Ordinary differential equations. London: Constable. Dover Publication (1955), New York.

Jacobson, N. (1953). Lectures in abstract algebra. Vol. II. New York: Van Nostrand.

Karlin, S. and J. McGregor. (1964). On some stochastic models in genetics. Stochastic Model in Medicine and Biology, edited by J. Gurland, pp. 245-279. Madison, Wisconsin: University of Wisconsin Press.

Karlin, S. and J. McGregor. (1972). Addendum to a paper of W. J. Ewens. Theoret. Popul. Biol. 3: 113-116.

Kato, T. (1966). Perturbation theory for linear operators. Berlin-Heidelberg-New York: Springer-Verlag.

Kimura, M. (1954). Process leading to quasi-fixation of genes in natural populations due to random fluctuation of selection intensities. Genetics 39: 280-295.

Kimura, M. (1955). Solution of a process of a random genetic drift with a continuous model. Proc. Nat. Acad. Sci. U. S. 41: 144-150.

Kimura, M. (1957). Some problems of stochastic processes in genetics. Ann. Math. Statist. 28: 882-901.

Kimura, M. (1962). On the probability of fixation of mutant genes in a population. Genetics 47: 713-719.

Kimura, M. (1968). Evolutionary rate at molecular level. Nature 217: 624-626.

Kimura, M.   (1968a).   Genetic variability maintained in a finite
    population due to mutational production of neutral and nearly
    neutral isoalleles. Genetical Res. 11: 247-269.

Kimura, M.   (1969). The number of heterozygous nucleotide sites
    maintained in a finite population due to steady flux of mutations.
    Genetics 61: 893-903.

Kimura, M.   (1970). The length of time required for a selectively
    neutral mutant to reach fixation through random frequency drift
    in a finite population. Genetical Res. 15: 131-134.

Kimura, M. and J. F. Crow.   (1964). The number of alleles that can
    be maintained in a finite population. Genetics 49: 725-738.

Kimura, M. and T. Maruyama.   (1969). The substitutional load in a
    finite population. Heredity 24: 101-141.

Kimura, M. and T. Maruyama.   (1971). Pattern of neutral polymorphism
    in a geographically structured population. Genetical Res. 18:
    125-131.

Kimura, M. and T. Ohta.   (1969). The average number of generations
    until fixation of a mutant gene in a finite population.
    Genetics 61: 763-771.

Kimura, M. and T. Ohta.   (1969a). The average number of generations
    until extinction of an individual mutant gene in a finite popula-
    tion. Genetics 63: 701-709.

Kimura, M. and T. Ohta.   (1973). The age of a neutral mutant persist-
    ing in a finite population. Genetics 75: 199-212.

Kimura, M. and G. H. Weiss.   (1964). The stepping stone model of
    population structure and the decrease of genetic correlation with
    distance. Genetics 49: 561-576.

Kolmogorov, A. N. and S. V. Fomin.   (1961). Elements of the theory of
    functions and functional analysis. Vol. 2. Albany, New York:
    Graylock Press. Translated from the first (1960) Russian edition
    by H. Kamel and H. Komm.

Li, W-H.   (1975). Total number of individuals affected by deleterious
    mutations in a finite population. Ann. Human Genet. 38:
    333-340.

Li, W-H. and M. Nei.   (1972). Total number of individuals affected
    by a single deleterious mutation in a finite population. Am. J.
    Human Genet. 24: 667-679.

Li, W-H. and M. Nei.   (1975). Drift variance of heterozygosity and
    genetic distance in transient states. Genetical Res. 25:
    229-248.

Malécot, G. (1948). Les mathématiques de l'hérédité. Paris: Masson et Cie. Mathematics of heredity (English translation). San Francisco: Freeman.

Malécot, G. (1967). Identical loci and relationship. Proc. 5th Berkeley Simp. Math. Statist. Probab. IV, pp. 317-332. Berkelay: University of California Press.

Malécot, G. (1975). Heterozygosity and relationship in regularly subdivided populations. Theoret. Popul. Biol. 8: 212-241.

Maruyama, T. (1970). Effective number of alleles in a subdivided population. Theoret. Popul. Biol. 1: 273-306.

Maruyama, T. (1970a). Stepping stone models of finite length. Advan. Appl. Probab. 2: 229-258.

Maruyama, T. (1970b). On the rate of decrease of heterozygosity in circular stepping stone models. Theoret. Popul. Biol. 1: 101-119.

Maruyama, T. (1970c). On the fixation probability of mutant genes in a subdivided population. Genetical Res. 15: 221-226.

Maruyama, T. (1971). The rate of decrease of heterozygosity in a population occupying a circular or a linear habitat. Genetics 67: 437-454.

Maruyama, T. (1971a). Speed of gene substitution in a geographically structured population. Am. Naturalist 105: 253-265.

Maruyama, T. (1971b). An invariant property of a structured population. Genetical Res. 18: 81-84.

Maruyama, T. (1972). Some invariant properties of a geographically structured finite population: Distribution of heterozygotes under irreversible mutation. Genetical Res. 20: 141-149.

Maruyama, T. (1972a). Rate of decrease of genetic variability in a two-dimensional continuous population of finite size. Genetics 70: 639-651.

Maruyama, T. (1972b). Distribution of gene frequencies in a geographically structured finite population. I. Distribution of neutral genes and of genes with small effect. Ann. Human Genet. 35: 411-423.

Maruyama, T. (1974). The age of an allele in a finite population. Genetical Res. 23: 137-143.

Maruyama, T. (1974a). A simple proof that certain quantities are independent of the geographical structure of population. Theoret. Popul. Biol. 5: 148-154.

Maruyama, T. and M. Kimura. (1971). Some methods for treating
continuous stochastic processes in population genetics. Japan.
J. Genetics 46: 407-410.

Maruyama, T. and M. Kimura. (1974). A note on the speed of gene
frequency changes in reverse directions in a finite population.
Evolution 28: 161-163.

Maruyama, T. and M. Kimura. (1975). Moments for sum of an arbitrary
function of gene frequency along a stochastic path of gene
frequency change. Proc. Nat. Acad. Sci. U. S. 72: 1602-1604.

Mather, K. (1969). Selection through competition. Heredity 24:
529-540.

Miller, G. F. (1962). The evolution of eigenvalues of a differential
equation arising in a problem in genetics. Proc. Cambridge Phil.
Soc. 58: 588-593.

Moran, P. A. P. (1958). Random processes in genetics. Proc.
Cambridge Phil. Soc. 54: 60-72.

Moran, P. A. P. (1960). The survival of a mutant gene under selection.
J. Australian Math. Soc. 1: 121-126.

Moran, P. A. P. (1962). The statistical processes of evolutionary
theory. Oxford: Clarendon Press.

Morse, P. M. and H. Feshbach. (1953). Methods of theoretical physics.
Part I and II. New York: McGraw-Hill.

Nagylaki, T. (1974). The moments of stochastic integrals and the
distribution of sojourn times. Proc. Nat. Acad. Sci. U. S. 71:
746-749.

Nagylaki, T. (1974a). The decay of genetic variability in geographically structured populations. Proc. Nat. Acad. Sci. U. S. 71:
2932-2936.

Nagylaki, T. (1977). The decay of genetic variability in geographically structured populations. III. Proc. Nat. Acad. Sci. U. S.
(in press).

Nei, M. (1968). The frequency distribution of lethal chromosomes in
finite populations. Proc. Nat. Acad. Sci. U. S. 60: 517-524.

Nei, M. (1971). Total number of individuals affected by a single
deleterious mutation in large populations. Theoret. Popul. Biol.
2: 426-430.

Nei, M. (1975). Molecular population genetics and evolution.
Amsterdam: North-Holland.

Nei, M. (1977). Am. J. Human Genet. (in press).

Nei, M. and S. Yokoyama. (1976). Effects of random fluctuation of
selection intensity on genetic variability in a finite population.
Japan. J. Genetics 51: 355-369.

Ohta, T. and M. Kimura. (1973). A model of mutation appropriate to estimate the number of electrophoretically detectable alleles in a finite population. Genetical Res. 22: 201-204.

Ray, D. (1956). Stationary Markov processes with continuous paths. Trans. Am. Math. Soc. 82: 452-493.

Robertson, A. (1964). The effect of non-random mating within inbred lines on the rate of inbreeding. Genetical Res. 5: 164-167.

Sato, K. (1976). A class of Markov chains related to selection in population genetics. J. Mathematical Society of Japan 28: 621-637.

Stewart, F. M. (1976). Variability in the amount of heterozygosity maintained by neutral mutations. Theoret. Popul. Biol. 9: 188-201.

Szegö, G. (1967). Orthogonal polynomials. Providence: American Mathematical Society.

Titchmarsh, E. C. (1946). Eigenfunction expansions associated with second-order differential equations. Oxford: Clarendon Press, 2nd edition (1962).

Trotter, H. F. (1958). Approximation of semi-groups operators. Pacific J. Mathemetics 8: 887-919.

Wasow, W. (1965). Asymptotic expansions for ordinary differential equations. New York: Interscience.

Watterson, G. A. (1962). Some theoretical aspects of diffusion theory in population genetics. Ann. Math. Statist. 33: 939-957.

Wehrhahn, C. F. (1975). The evolution of selectively similar electrophoretically detectable alleles in finite natural populations. Genetics 80: 375-394.

Whittaker, E. T. and G. N. Watson. (1927). A course of modern analysis. Cambridge: Cambridge University Press.

Wright, S. (1931). Evolution in Mendelian populations. Genetics 16: 97-159.

Wright, S. (1939). The distribution of self-sterility alleles in populations. Genetics 24: 538-552.

Wright, S. (1948). On the roles of directed and random changes in gene frequency in the genetics of populations. Evolution 2: 279-294.

Wright, S. (1948a). Genetics of populations. Encyclopedia Britannica 10: 111, 111A-D, 112.

Wright, S. (1965). The distribution of self-inconpatibility alleles in populations. Evolution 18: 609-619.

Wright, S. (1969). Evolution and genetics of populations. Vol. 2. The theory of gene frequencies. Chicago: University of Chicago Press.

Biomathematics

Springer-Verlag
Berlin
Heidelberg
New York

This series aims to report new developments in biomathematics research and teaching – quickly, informally and at a high level. The type of material considered for publication includes:

1. Preliminary drafts of original papers and monographs

2. Lectures on a new field, or presenting a new angle on a classical field

3. Seminar work-outs

4. Reports of meetings, provided they are

    a) of exceptional interest and

    b) devoted to a single topic.

Texts which are out of print but still in demand may also be considered if they fall within these categories.

The timeliness of a manuscript is more important than its form, which may be unfinished or tentative. Thus, in some instances, proofs may be merely outlined and results presented which have been or will later be published elsewhere. If possible, a subject index should be included. Publication of Lecture Notes is intended as a service to the international scientific community, in that a commercial publisher, Springer-Verlag, can offer a wider distribution to documents which would otherwise have a restricted readership. Once published and copyrighted, they can be documented in the scientific literature.

Manuscripts

Manuscripts should comprise not less than 100 and preferably not more than 500 pages.

They are reproduced by a photographic process and therefore must be typed with extreme care. Symbols not on the typewriter should be inserted by hand in indelible black ink. Corrections to the typescript should be made by pasting the amended text over the old one, or by obliterating errors with white correcting fluid. Authors receive 75 free copies and are free to use the material in other publications. The typescript is reduced slightly in size during reproduction; best results will not be obtained unless the text on any one page is kept within the overall limit of 18 x 26.5 cm (7 x 10½ inches). The publishers will be pleased to supply on request special stationery with the typing area outlined.

Manuscripts in English, German or French should be sent to Dr. Simon Levin, Center for Applied Mathematics, Olin Hall, Cornell University Ithaca, NY 14850/USA or directly to Springer-Verlag Heidelberg.

Springer-Verlag, Heidelberger Platz 3, D-1000 Berlin 33
Springer-Verlag, Neuenheimer Landstraße 28–30, D-6900 Heidelberg 1
Springer-Verlag, 175 Fifth Avenue, New York, NY 10010/USA

ISBN 3-540-08349-9
ISBN 0-387-08349-9